T0358333

Cambridge Elements ≡

Elements in International Relations
edited by
Jon C. W. Pevehouse
University of Wisconsin–Madison
Tanja A. Börzel
Freie Universität Berlin
Edward D. Mansfield
University of Pennsylvania
Associate editor – International Security
Anna Leander
Graduate Institute Geneva

DRONES, FORCE AND LAW

European Perspectives

David Hastings Dunn
University of Birmingham
Nicholas J. Wheeler
University of Birmingham
with
Jack Davies
University of Leiden
Zeenat Sabur
University of Manchester

CAMBRIDGE
UNIVERSITY PRESS

Shaftesbury Road, Cambridge CB2 8EA, United Kingdom

One Liberty Plaza, 20th Floor, New York, NY 10006, USA

477 Williamstown Road, Port Melbourne, VIC 3207, Australia

314–321, 3rd Floor, Plot 3, Splendor Forum, Jasola District Centre, New Delhi – 110025, India

103 Penang Road, #05–06/07, Visioncrest Commercial, Singapore 238467

Cambridge University Press is part of Cambridge University Press & Assessment, a department of the University of Cambridge.

We share the University's mission to contribute to society through the pursuit of education, learning and research at the highest international levels of excellence.

www.cambridge.org
Information on this title: www.cambridge.org/9781009451529

DOI: 10.1017/9781009451499

© David Hastings Dunn, Nicholas J. Wheeler, Jack Davies, and Zeenat Sabur 2023

This publication is in copyright. Subject to statutory exception and to the provisions of relevant collective licensing agreements, no reproduction of any part may take place without the written permission of Cambridge University Press & Assessment.

First published 2023

A catalogue record for this publication is available from the British Library

ISBN 978-1-009-45152-9 Hardback
ISBN 978-1-009-45148-2 Paperback
ISSN 2515-706X (online)
ISSN 2515-7302 (print)

Cambridge University Press & Assessment has no responsibility for the persistence or accuracy of URLs for external or third-party internet websites referred to in this publication and does not guarantee that any content on such websites is, or will remain, accurate or appropriate.

Drones, Force and Law

European Perspectives

Elements in International Relations

DOI: 10.1017/9781009451499
First published online: December 2023

David Hastings Dunn
University of Birmingham

Nicholas J. Wheeler
University of Birmingham

with
Jack Davies
University of Leiden

Zeenat Sabur
University of Manchester

Author for correspondence: David Hastings Dunn, d.h.dunn@bham.ac.uk

Abstract: The central argument set out in this Element is that the combination of a perceived radical change in the threat environment post 9/11, and the new capabilities afforded by the long silent reach of the drone, have put pressure on the previously accepted legal frameworks justifying the use of force. This has resulted in disagreements – both articulated and unarticulated – in how the Western allies should respond to both the legal and operational innovations in the use of force that drones have catalysed. The Element focuses on the responses of the UK, France, and Germany to these developments in the context of the changing US approach to the use of force. Locating itself at the interface of international law and politics, this is the first attempt to look at the interplay between technological innovations, legal justifications, and inter-alliance politics in the context of the use of armed drones.

Keywords: drones, force, international law, counterterrorism, European

© David Hastings Dunn, Nicholas J. Wheeler, Jack Davies, and Zeenat Sabur
2023

ISBNs: 9781009451529 (HB), 9781009451482 (PB), 9781009451499 (OC)
ISSNs: 2515-706X (online), 2515-7302 (print)

Contents

Introduction

The central focus of *Drones, Force and Law* is that the post 9/11 security environment has led to contestation among key European states as to the acceptable limits of the use of force for counterterrorist purposes. The fissures that have opened up between the UK government on the one hand, and the German and French governments on the other have been in response to the highly controversial legal justifications the US and UK governments have employed to legitimate using force beyond recognised battlefields. The US government has been the primary political and legal innovator of this use of force since 9/11 – epitomised by its reliance on the technology of the armed drone. But this US legal innovation has led subsequently to others, notably the UK government, also employing what has been such a controversial legal rationale. The key argument of this Element is that in this new context, international law itself has become a battleground over the extent to which legal rules can constrain these uses of force or whether the law is being used to enable them. The consequence of these developments is that there is no longer a consensus among the three major European military powers – the primary focus of this Element – on what constitutes the limits on the legal use of force beyond the battlefield.[1] We argue that this lack of consensus has implications for the regulation of force in international society and caution against the dangers of pushing out the boundaries of the legitimate use of force in ways that weaken the international legal order.

Critics of the new uses of force for counterterrorist purposes argue that these actions break existing international law.[2] By contrast, the governments employing force in this way argue that such uses, whether by drones or other means, are in conformity with existing international law. Governments seek to legitimise their own uses of force by reference to the law while delegitimising the actions of their opponents. But what happens when legal interpretations clash, as with the use of armed drones for counterterrorist purposes beyond recognised

---

[1] This Element defines a non-battlefield targeted killing as one that takes place where a state has no armed forces directly involved in an armed conflict. It is a military attack that is aimed at suspected terrorists or other combatants inside the borders of another state where that state has not provided consent and where there is no express UN Security Council authorisation (see also MacDonald, 2017; Banka and Quinn, 2018 for a discussion of definitional issues around the idea of targeted killing).

[2] Given the limits of space, the Element focuses on the *jus ad bellum*. This is defined by the International Committee of the Red Cross as 'the conditions under which States may resort to war or to the use of armed force in general', www.icrc.org/en/publication/0703-international-humanitarian-law-answers-your-questions. The military conduct of states in an armed conflict (*jus in bello*) and the body of International Humanitarian Law (IHL) regulating force in an armed conflict as well as the question of the applicability of International Human Rights Law (IHRL) to govern the conduct of military operations is not the focus of this Element.

battlefields? Both sides in the debate use the same legal materials to justify their positions, but each side claims on the basis of their interpretation of these materials that a particular use of force is either lawful or unlawful depending on their purposes. Locating itself at the interface of the disciplines of International Relations (IR) and International Law, this Element goes beyond this argument over legality/illegality by arguing that states use international law as a tool of strategising (Simpson, 2000; Wheeler, 2004; Peevers, 2013; Hurd, 2017a, 2017b; Schmidt and Trenta, 2018; Birdsall, 2022).

Building on the existing literature, the Element develops a novel conceptualisation of what we call legal strategising, the concept is developed more fully later in the Introduction, which we explain in terms of the process by which policymakers and their legal advisors engage in legal justification and argumentation to legitimate particular courses of military action.[3] The Element conceptualises legal strategising in terms of three competing approaches which we call *innovation*, *conventionalism*, and *exceptionalism*. In doing so, we have two purposes. The first is to map the complex interplay and contestation between innovation, conventionalism, and exceptionalism in the legal justifications proffered by the UK, French, and German governments in relation to the use of armed drones for counterterrorist purposes. Second, to explain the divergences in the legal justifications employed by the UK, French, and German governments. The central motivating puzzle guiding this Element is that despite drone technology being available to all three European states and a common threat perception with regard to non-state terrorist actors (in particular, Al-Qaeda (AQ), Al-Qaeda in the Arabian Peninsula (AQAP), and the Islamic State in the Levant (ISIL)), the UK, French, and German governments have responded very differently in terms of their own acquisition and use of armed drone technology and their legal and political response to others – inside and outside Europe – who have used it for non-battlefield targeted killing.

What the empirical sections reveal is that these differences in legal strategising can be explained in terms of four variables. These are (i) domestic politics; (ii) alliance politics; (iii) individual leadership; and (iv) precedent-setting. It is the different operation of these four variables across the three cases that explains the divergence in legal strategising in the British, French, and German cases. These four variables have been identified through an inductive research enquiry where we have asked the same questions in relation to each case. These are:

---

[3] There are similarities between our concept of legal strategising and the idea of 'Lawfare'. The latter has been defined by Charles Dunlap as 'the use of law as a weapon of war' or 'a means of realizing a military objective' (2001: 2, 4; see also Kittrie, 2016). Our focus is not on the use of law as a weapon in military conflict, but as a conceptual tool to interrogate the legal justifications of states that engage in and respond to the use of force for non-battlefield targeted killing.

(i) what kinds of legal justifications have been adduced for the use of force in non-battlefield targeted killing, including the use of armed drones where such force has been employed; (ii) what explains the selection of these legal strategies and how far have they been contested domestically within the states concerned; (iii) how far have policymakers and their legal advisors sought to engage in law-making when advancing new justifications for the use of force; and (iv) how far have policymakers and their legal advisors acquiesced or legitimated in the use of force by others, and what explains this? The Conclusion to the Element uses this framing to explain in relation to each case the divergences in legal strategising.

These questions are generic ones in relation to the use of force, but they have been given added urgency and significance because of the availability and utilisation of armed drone technology on the part of European states. It is to the enabling condition of this technology that we now turn.

The Technological Push of Drone Technology

The term 'drone' has recently taken on a wider definition in part due to the war in Ukraine where the designation has come to be associated with any novel use of an uncrewed flying instrument, be that loitering munitions or preprogrammed Iranian 'drones' or the more advanced Turkish-made TB2. By contrast we define a drone here as a *remotely piloted crewless aircraft*. In using this definition, we are explicitly excluding Lethal Autonomous Weapon Systems (LAWS) as they are not the subject of the legal arguments and contestations discussed in this Element.

Drone technology has provided a novel means through which to prosecute the Global War on Terror (GWOT). The conflict which began in the aftermath of the 9/11 attacks went on to become a global military counterterrorism campaign, targeting disparate and loosely-connected terrorist groups and networks, often in remote geographies. Key to this evolution was the expanded ability to project force at remote distances provided by drones, which facilitated the pivot from the strategically limited large-scale boots-on-the-ground deployments that characterised the early, more conventional stages of the GWOT, to the light-footprint *modus operandi* of remote warfare that became the preferred method of the campaign in its later years. For governments, drones represent the promise that their delivery of lethal force can be done with great precision and limited risk, an ideal pairing for publics in liberal-democratic states, which increasingly expect their wars to be bloodless for both combatants and civilians (Coker, 2001; Kaag and Kreps, 2014; Zehfuss, 2018). The precision strike technology of the drone and its capacity for persistent air surveillance offer the seductive, though disputed, possibility of using force in ways that minimises harm to non-combatants on an unprecedented scale (Gregory, 2017). For some, they represent the apogee of the trend towards increasing precision that developed through the Revolution in Military Affairs in the 1990s. The employment of drone technology avoids the political cost of having to guard against both casualties and casualty-aversion. For their critics, this reduction of political risk and profile has led to a situation where the sheer volume of attacks indicates an approach to target identification that is more permissive of the use of force.

For some writers the shift towards remotely controlled systems represents a categorical departure into something wholly new (Kreps and Lushenko, 2023). The focus of this study on drones, as opposed to other forms of direct fire weapon systems, recognises that this new capability creates opportunities for military strikes in circumstances that are legally contested. Drones have opened up a new form of aerial warfare. Whereas previously governments would not risk the loss of aircrew in contested circumstances due to the political consequences which would follow, drone warfare reverses this logic (Gusterson, 2016; see also Kreps, 2016).

The long endurance of drones enabling extended loitering over potential targets, paired with the integration of both Intelligence, Surveillance and Reconnaissance (ISR) and real-time lethal strike capabilities into one uncrewed platform opens up the possibility of persistent surveillance supported by rapid strike options. Armed drones, especially when operated in shift patterns, offer decision-makers the possibility of near-perpetual ISR and strike options even in extremely remote contexts. As precision military instruments, drone strikes complement the much more extensive use of Special Forces. But drones can be even more appealing to decision-makers due to the fact that their use can be simultaneously secret, invisible, anonymous, and deniable (for both parties).

Set against the claims that drones represent a step-change in the nature of warfare, some observers argue that drones offer no more than an extension of pre-existing trends in the use of crewed aircraft and are best understood as expansions of existing capabilities (Boyle, 2020). This argument has some validity 'where drone operators track their own troops as well as insurgents', what Hugh Gusterson categorises as 'mixed drone warfare'(Gusterson, 2016: 15). By contrast, drone use is much more contested in what Gusterson describes as 'pure drone warfare'. For him, this is a situation where drones operate solely to track and kill suspected terrorists, outside the 'legal framework that applies to an internationally recognised warzone' (Gusterson, 2016: 15). This definition overlaps with our categorisation and focus on 'non-battlefield targeted killing' but where Gusterson views such strikes as outside the remit of the law, this Element investigates the contestation over the legality of such strikes.

We are not arguing that the advent of drone technology alone has determined these new possibilities of action (see also Moyn, 2021). As the Element shows, other military means have also been employed for counterterrorist purposes. But with this option available to them, decision-makers have the institutional-ised capacity to employ force in ways which were previously not available to them because they are capable of crossing sovereign borders with impunity in order to conduct both ISR and strike operations. The use of drones challenges traditional geographical distinctions between the battlefield and non-battlefield. In the post 9/11 environment it was the US government that first embraced the opportunities of using armed drones outside of recognised battlefields. Used in conjunction with Special Forces, drones were initially employed in Yemen and Pakistan, but then their use spread geographically to include Somalia, Libya, Nigeria, and Syria among others.

What is important about these new US uses of force was not just the material deployment of US power, but the novel legal justifications that were employed to legitimise them. A week after the 9/11 attacks, the US Congress passed the 2001 Authorization for Use of Military Force Act (AUMF). This empowers the President to 'use all necessary and appropriate force against those nations, organizations, or persons he determines planned, authorized, committed, or aided the terrorist attacks'.[4] The AUMF, a domestic legal instrument, was aimed at securing legitimation and enabled a lowering of the threshold for the US government's use of force against global terrorist threats.

Alongside this act of domestic legal innovation, the US government has adopted two highly controversial interpretations of the law of self-defence to justify its non-battlefield targeted killing of suspected terrorists. The first and most prominent is the reinterpretation of the meaning of imminence in the customary law of self-defence.[5] The traditional meaning of imminence derives from the *Caroline* case of 1837. The US and UK governments were in dispute as a result of the latter launching an armed expedition that destroyed a US vessel, *Caroline,* that was preparing to transport guerrilla forces to assist Canadian

---

[4] US Congress (2001). '*Public Law 107–40, Joint Resolution To authorize the use of United States Armed Forces against those responsible for the recent attacks launched against the United States'.* www.congress.gov/107/plaws/publ40/PLAW-107publ40.pdf.

[5] Customary international law is one of the main sources of international law (the other being Treaties and general principles of law). Customary international law depends on state practice; states must engage in a practice that they believe is permitted or required by the law. However, it is essential that a significant number of states (there is no agreement among international lawyers as to what number is required here) recognise that the practice is permitted by the law. This is known as *opinio juris* and the International Court of Justice (ICJ) in its *North Sea Continental Shelf* Judgment defined this as 'a belief that [a] practice is rendered obligatory by the existence of a rule of law requiring it' (North Sea Continental Shelf Cases, Judgment of 20 February 1969, www.icj-cij.org/public/files/case-related/51/051–19690220-JUD-01–00-EN.pdf, see also Byers, 1999).

rebels who were challenging British colonial rule. In response to the UK government's claim that its action was justified on grounds of self-defence, US Secretary of State Daniel Webster argued in a now oft-repeated formulation that, 'It will be for [the British] government to show a necessity of self-defence, instant, overwhelming, leaving no choice of means, and no moment of deliberation' (Jennings, 1938). This traditional understanding of imminence has been challenged by successive US governments since 9/11 which have argued for a different, non-temporal conception of imminence that emphasises removing threats, even before they have fully manifested themselves in visible preparations for attack. Such a view stretches the interpretation of what counts as an imminent threat well beyond anything that Webster had in mind when he formulated what has become known as the *Caroline* standard.

The second controversial interpretation relates to what is known as the 'unable and unwilling doctrine'. This is claimed to give states that have been victims of attacks by non-state actors – or where such attacks are believed to be imminent – the legal right to use force on the territory of another state where that state is unable or unwilling to prevent or stop such attacks being conducted from their territory. This legal claim was originally formulated by the Nixon administration to defend its use of force in Cambodia, though was roundly rejected at the time (Heller, 2019; see also Stevenson, 1970 and for a critique of its legal basis, Mignot-Mahdavi, 2023: 92–7).

Both of these US legal interpretations of what is permissible under the law of self-defence in relation to the use of force in a counterterrorist role have proved highly controversial within US domestic debates (Murphy, 2005; Green, 2009). And as we will show in the rest of the Element, these have created a split among Washington's major European military allies over whether such interpretations go beyond the bounds of what can plausibly be justified as lawful self-defence. To explain the lack of consensus within intra-European relations over the boundaries governing the legal use of force, we develop a theoretical framework that is located in an English School conception of IR that views international law as both constraining and enabling of state actions.

Law in International Society

The theoretical starting point for this Element is the proposition that states form an international society amidst the condition of international anarchy (defined as the absence of a world or global government) that they dwell in. This fundamental assumption defines what has become known as the English School approach (Buzan, 2015 provides an excellent overview of the approach;

see also Bull, 1977; Jackson, 2000; Hurrell, 2007). For the English School, it is the production and reproduction of international society that holds the key to the provision of international order and the minimisation of violence between states that this order makes possible. Hedley Bull, one of the school's pioneering thinkers, defined the idea of international society as existing 'when a group of states, conscious of certain common interests and values, form a society in the sense that they conceive themselves to be bound by a set of common rules in their relations with one another, and share in the working of common institutions' (1977: 13). For Bull, what separates out an *international system* where states interact frequently, but do so without any element of sociality, from an *international society* is that state representatives recognise that they owe others an account of their behaviour in terms of the 'common rules' they all accept (Bull, 1977: 45). The most important body of rules for the English School is the practice of international law.

Bull argued that all societies at a minimum have to establish 'primary rules' for the maintenance of order, and that these foundational rules provide for security against violence, the stability of possessions, and the sanctity of promises (1977: 3–8). The defining characteristic of an international society is that states recognise 'laws which are binding ... in their relations with one another' (Bull, 1966: 53). In his 1977 opus, *The Anarchical Society*, he defined these legal obligations in terms of states being 'bound by certain rules in their dealings with one another, such as that they should respect one another's claims to independence, that they should honour agreements into which they enter, and that they should be subject to certain limitations in exercising force against one another' (Bull, 1977: 13). The idea of states being 'bound' by a common set of legal rules is the kernel of the English School's approach to international order.

Our key interest lies in how states use international law as a tool of legitimation in the contemporary society of states. Governments and their legal advisors do so by instrumentalising the shared legal rules to justify their actions, criticise those of their opponents and adversaries, and seek a new consensus where they are relying on novel interpretations of existing rules to justify actions that were not previously covered by that rule. A core English School proposition is that the legal rules that states recognise in their relations with one another are not endlessly manipulable and there is a constraining effect on state behaviour if actors cannot find plausible justifications for their actions. Bull explained it in the following terms: 'There are, of course, differences of opinion as to the interpretation of the rules and their application to concrete situations; but such rules are not infinitely malleable and do circumscribe the range of choice of states which seek to give pretexts in terms of them' (Bull, 1977: 45; see also Henkin, 1979; Franck, 1990, 2006; Wheeler, 2000).

Law is at the heart of strategies of legitimating the use of force because state leaders, civil servants, and diplomats appreciate that 'conceptions of rightful conduct' (a key component along with rightful conceptions of membership in Ian Clark's definition of legitimacy) depend upon validating their action in legal terms. In addition to the value of being seen to comply with the law internationally, Governments recognise the political value at home of being able to justify their actions as lawful, especially when it comes to the use of force. For example, Lushenko et al.'s quantitative empirical study of American and French public attitudes to armed drone strikes demonstrates that perceived compliance with international law – expressed in terms of United Nations (UN) authorisation – is the strongest predictor of public support for strikes in these countries (Lushenko et al., 2022b; see also Kreps, 2014; Kreps and Wallace, 2016).

The constraining power of international law is emphasised by the English School, constructivist theorists in IR, and those international lawyers who emphasise what Thomas Franck called 'The Power of Legitimacy' (1990, 2006; see also Claude, 1966). For Franck, the 'compliance pull' (2006: 93; see also 1990) of a legal rule depends on its legitimacy. Here, the most important element that produces and reproduces this legitimacy is the 'determinacy' of a rule because the more determinate a rule is, the more there is 'an ascertainable understanding of what it permits and what it prohibits. When that line becomes unascertainable, states are unlikely to defer opportunities for self-gratification' (Franck, 2006: 93). The notion of law as a constraining factor in policymaking does not imply any physical or material restraint on state actions; rather, the idea of constraint employed in this Element is derived from an understanding of how state actors and their domestic publics are embedded within a normative context structured by legal rules. Abraham and Antonio Handler Chayes expressed this position in the following terms: 'It is almost always an adequate explanation that for an action, at least prima facie, that it follows the legal rule . . . . It is almost always a ground for disapproval that an action violates the norms' (Chayes and Chayes, 1998: 119). Justifications that are couched in terms of particular interests or value, and which rely on an appeal to idiosyncratic forms of reasoning, or which are viewed as treating the legal framework in a cavalier fashion, fail to meet this test.

Even when policymakers know that they are breaking established interpretations of the law, they rarely admit this publicly, preferring to offer a legal justification, however strained and implausible, that is in conformity with the rules. If a state openly admitted that it was violating the law, giving a justification for its conduct only in terms of that state's values and beliefs, then it would be treating others with contempt and would in Bull's words 'place in jeopardy all the settled expectations that states have about one another's behaviour' (Bull, 1977: 45).[6] States accept the legitimacy of the law because they recognise this as an obligation that flows from being a responsible member of international society (Franck, 1990; Clark, 2005; Hurrell, 2007). It follows that if governments recognise an obligation to justify actions in terms of shared legal rules and principles, legal arguments cannot be endlessly manipulated if actors perceive the need to secure legitimation from other members of international society (Johnstone, 2003; Reus-Smit, 2004). Russian President Vladimir Putin came to appreciate this after his invasion of Ukraine in February 2022. Supporting Bull's claim that states do not openly admit to breaking the law, Russia sought to justify its use of force against Ukraine on the grounds that it was invited in by the 'independent' republics of Donetsk and Luhansk. Given that only Russia, Belarus, Syria, Nicragua, North Korea, Venezuela, Sudan, and the Central African Republic recognised these territories as independent sovereigns, such a justification clearly failed to pass what Franck called 'the laugh test' (2006: 96). But the fact that the Russian government felt it necessary to engage in such a tortuous justification which was so obviously false shows that Russia recognised the binding legal obligation in Article 2(4) of the UN Charter which prohibits the use of force against 'the territorial integrity or political independence of any state'. Russia's legal justification has been widely repudiated – though not universally (see the Conclusion) – by international society and has led to many states imposing significant political and economic costs on Russia for what is seen by them as its flagrant breach of the rules.

The conception of law as constraining on state actions has to be set against the realist argument that law can also be enabling, especially in relation to the use of force. The constraining effect is challenged by scholars and practitioners of this persuasion because international law is not an uncontested corpus of rules and that what counts as compliance in any particular case is open to contestation and legal disputation. A good example is the dispute over the legality of the 2003 US-led invasion of Iraq. The US, UK, and Australian governments that spearheaded the use of force claimed that their military intervention was legal, despite this use of force lacking specific UN Security Council authorisation. By contrast, other states, notably Russia, China, and India, but also including key allies like France and Germany, argued that the intervening states were breaking core UN Charter principles of sovereignty, non-intervention, and non-use of force enshrined in Articles 2(7) and 2(4), respectively.

---

[6] Bull argued that 'the offending state usually goes out of its way to demonstrate that it still considers itself (and other states) bound by the rule in question' (Bull, 1977: 138).

In an international realm that lacks an authoritative sovereign to interpret and adjudicate the application of the law in particular cases, realism argues that power determines which legal interpretations are privileged. *Contra* Franck's claim that there are determinate and constraining legal rules governing the use of force that states have a long-term interest in upholding (1990, 2006), Richard Betts contends that law has never exerted a constraining influence on state decisions to use force, with legal justifications 'figuring mainly as rationalizations . . . for decisions made on other grounds' (Betts, 2003; see also Hurd, 2017a). But this does not make international law window dressing for realists; instead law becomes yet another battleground in the unending struggle for power. As Hans J. Morgenthau, one of the founding fathers of political realism claimed, the powerful are always seeking 'to shake off the restraining influence that international law might have upon their foreign policies, *to use international law instead for the promotion of their national interests, and to evade legal obligations that might be harmful to them*' (Morgenthau, 1967: 268–9).

Investigating the relationship between law and politics as both enabling and constraining of the use of force is the point of departure of this Element. We follow those international lawyers and IR theorists who argue that international law always operates in varying degrees of indeterminacy where actors seek to use the law for their political purposes. Martti Koskenniemi provided a strong formulation of this position when he wrote, 'The choice is not between law and politics, but between one politics of law, and another' (2005: 123). A key contribution to this way of thinking is Ian Hurd's *How to Do Things with International Law* (2017a) which exposes the limitations of trying to separate out law and politics. Instead, he argued that 'international law cannot be separated from state practice, and state practice does not exist independent of the legal explanations, justifications, and rationalizations that governments give for it' (2017, especially 7, 131; see also Simpson, 2004; Wheeler, 2004; Peevers, 2013). The use of these legal arguments and counter-arguments – and the clash of interpretations that follow – is what we understand as *legal strategising.* We define this but *the articulation of a state's justification of its interpretation of domestic and international law congruent with its political interests and national values, which is derived from a process of internal deliberation in which policy-makers and their legal advisors argue over whether certain courses of action can be plausibly justified within the confines of domestic and international law.* The Element develops a typology: *innovating, conventionalist,* and *exceptionalist* (set out below) of legal strategising that reflects the interplay between enabling and constraining conceptions of the law.

Innovating

Governments employ an 'innovating' legal strategy when they seek to advance new legal interpretations of existing rules and principles to justify uses of force that would previously have been excluded as illegal under existing dominant interpretations of those rules. The US government's legal strategising after 9/11 in relation to the meaning of imminence in the customary law of self-defence is a textbook example of an innovating legal strategy at work. We will show in Section 1 how the UK government employed a similar innovating legal strategy to justify its use of force against Reyaad Khan, a UK national operating in Syria in August 2015. An innovating legal strategy has a strong affinity with what the political theorist Quentin Skinner called the 'innovating ideologist'. Skinner argued that all actors perceive a need to legitimate their actions and that conceptual change is likely to be most successful when actors are able to justify new claims in terms of established and accepted normative frameworks. He maintained that the 'innovating ideologist' seeks to manipulate existing concepts to legitimate actions that would previously have been excluded by it (Skinner, 1974, 1988; see also Bentley, 2014; Trenta, 2018).

The innovating approach to legal justification is rooted in a view of the law as enabling rather than constraining, but such innovations are not always successful and often lead to major pushback from governments, civil society groups, and international lawyers that seek to affirm the importance of existing legal interpretations of the rules – we call this 'conventionalist' legal strategising. We show in the sections to follow that the response of many European governments, human rights non-governmental organisations (NGOs), parliamentarians, and international lawyers has been to challenge the new legal claims justifying a more expansive interpretation of imminence.

Conventionalist

Once actors take on the burden of legal justification, they enter into a realm of reasoning in which actions must be justified in terms of shared normative precepts expressed in legal terms. Moreover, if actions cannot plausibly be justified in these terms, then actors risk being robbed of political legitimacy. Skinner's recognition of the role played by innovating idealists in the process of conceptual change was matched by an understanding of the limits to such innovative behaviour. In a convergent argument with Bull and the English School and international lawyers like Franck (and earlier Henkin) and Chayes and Chayes, Skinner wrote that, 'the agent cannot hope to stretch the application of the existing principles indefinitely; correspondingly, [the agent] can only hope to legitimate a restricted range of actions … any course of action is inhibited from occurring if it cannot be legitimated' (Skinner, 1988: 117). As a result, governments seeking to maximise the perceived legitimacy of their actions will generally rely on those legal justifications whose validity is established and largely unquestioned. Policymakers and their legal advisors employ a conventionalist legal strategy when they constrain their uses of force to conform to the shared legal precepts of international society in the expectation that others will be similarly constrained.

Exceptionalist

The third type of legal strategising discussed in the Element is the exceptionalist one. This has both a legal and a moral component. There is a literature on legal exceptionalism (Scheuerman, 1999; Byers, 2003; Simpson, 2004; Clark, 2011) which refers to how the great powers have historically carved out legal exceptions to the established rules that have been legitimated, albeit to varying degrees, by international society. Appeals to exceptionalist legal claims usually stem from the belief that the existing legal framework is inadequate to meet contemporary challenges tempered by a recognition that if *all* states were to claim and exercise the legal right in question, this would jeopardise the existing legal framework in a way that would weaken the international order (Scheuerman, 1999; Simpson, 2004; Clark, 2011).

The Bush Doctrine set out in President Bush's West Point speech of March 2002 of using force to prevent 'rogue states' in alliance with transnational terrorist networks from acquiring Weapons of Mass Destruction (WMD) would have been an opportunity for the US government to claim a legal exception to the rules governing the use of force on the grounds that the United States faced an exceptional situation of vulnerability after 9/11 (Hendrickson, 2002; Byers, 2003; Wheeler, 2003). However, what is striking is that at no point did the US government claim such a legal exception when it justified the use of force against Iraq in 2003. Instead, it relied on existing conventionalist legal justifications, namely, the Iraqi government's failure to comply with existing UN Security Council Resolutions relating to Iraq's disarmament of WMD as required after Saddam Hussein's defeat in the 1991 Gulf War.

Nor has there been any attempt by subsequent presidents to carve out special legal claims relating to the US reliance on a non-temporal doctrine of imminence. Equally, as we show in the following section, the UK government has not sought to justify its non-battlefield targeted killings in terms of special legal rights. Instead, like its US counterpart, has adopted an innovating legal strategy in relation to the law of self-defence that potentially could be invoked by others (this is discussed further in the Conclusion).

The other type of exceptionalist legal strategy in our framework is one which eschews legal justification altogether and relies exclusively on moral claims to justify the use of force. Actors who invoke exceptionalist moral arguments to justify their uses of force, whether for national security or humanitarian imperatives, can be seen as supporting the legitimacy of conventionalist interpretations of the law through their explicit demarcation of the acceptable limits of its use. A good illustration of this type of moral exceptionalism is the legal justifications employed by NATO states for their use of force in Kosovo in March 1999. With the important exception of the UK government that did offer an explicit legal defence of its actions (Wheeler, 2000), the other then eighteen members of the alliance defended their use of force on a mix of moral and political reasons that compelled military action. They did this because all the other available legal rationales were rejected as either failing the test of legal plausibility or they risked setting a dangerous precedent. A logical extension of this approach to the use of force for non-battlefield targeted killings, and one, as we explore in Section 2, that has been advocated in French strategic reasoning, is for policymakers to justify a use of force that cannot be justified under the existing legal rules by claiming that this is necessary on grounds of urgent national security considerations.

The richness and value of seeing the law as being in constant tension between enabling (innovating) and constraining (conventionalist) approaches is that it highlights how governments can use interpretive controversies over the meaning of the law to suit their purposes. This could be seen at work in the Trump Administration's legal justifications for its use of an armed drone in January 2020 to kill General Soleimani, a high-ranking official of the Iranian state. In writing to the UN Security Council to meet the US reporting requirement under Article 51 of the Charter, the US Permanent Representative explained that the US government had carried out the strike 'in the exercise of its inherent right of self-defense'.[7] In his interview on *Fox and Friends*, then US Secretary of State Mike Pompeo explained that Soleimani had 'got hundreds of American lives' blood on his hands. But what was sitting before us was his travels throughout the region and his efforts to make a significant strike against Americans'.[8] On 3 January, Trump told reporters at Mar-a-Lago that Soleimani 'was plotting imminent and sinister attacks on American diplomats and military personnel, but we caught him in the act and terminated him'.[9]

---

[7] The letter was signed by US Permanent Representative to the UN Security Council Ambassador Kelly Craft on 9 January 2020 (S/2020/20), https://undocs.org/pdf?symbol=en/S/2020/20.

[8] Pompeo, M. 2020a. Secretary of State Pompeo interviewed on Fox and Friends, US Department of State,www.state.gov/secretary-michael-r-pompeo-with-steve-doocy-ainsley-earhardt-and-brian-kilmeade-of-fox-and-friends/.

[9] US President, Donald J Trump. 2020. Remarks by President Trump on the Killing of Qassem Soliemani, 3 January, https://ge.usembassy.gov/remarks-by-president-trump-on-the-killing-of-qasem-soleimani-january-3/.

Despite these justifications, the administration was subsequently criticised by a number of legal and political commentaries on the attack (O'Connell, 2020; Milanovic, 2020). For many observers, the Trump administration had failed to provide convincing evidence that an attack was imminent. In levying this criticism, these observers adopt a temporal definition of imminence whereas the administration had moved on to operating with a non-temporal one.

The Solemani case illustrates well the limits of a rules-based approach to international law where actions can be judged to be simply either legal or illegal. Instead, what the Solemani case shows is how governments strive to find legal cover, even when they know that this challenges established legal precepts. It is then what Franck called the 'jurying' function of UN political organs, crucially the Security Council, other governments, and wider world public opinion to decide the validity of these competing legal claims in the context of their actions (2002: 186; see also Wheeler, 2004: 41–8).

A key contribution of this Element lies in showing how this enabling and constraining tension has played out in the legal strategising of the major European military powers as they have responded to the innovating US legal justifications for its uses of armed drones in a post 9/11 world.

Case Studies

The puzzle that guides the research is what explains the divergence in legal strategising between the French, German, and UK governments after 2015 when all three have held the same threat perceptions and had the same access to armed drone technology. European reactions to the altered security environment following the US response to the 9/11 attacks have varied and developed over time. This has resulted in disagreements – both articulated and unarticulated – in how the Western allies should respond to both the legal and operational arguments and arguments and innovations in the use of force that drones have provided.

We have selected the cases of the United Kingdom, France, and Germany purposively on the basis of the following criteria: first, there is variation in the legal justifications employed by each of these states for, and in response to, the use of force, including armed drones, for counterterrorist purposes. Second, these three states are the most capable military allies of the United States within NATO; and hence are in the best position to project force internationally. Third, as members of the NATO alliance they have all pledged to come to the military assistance of the United States as a result of the invocation of Article V of the NATO Treaty after the September 2001 terrorist attacks. Finally, each country has a different and changing level of armed drone capability in the post 9/11 period.

The United Kingdom is a particularly interesting case because it is the UK government's killing of Khan in August 2015 and its public justification of this action through a broadened interpretation of imminence that led to a rupture in the consensus that had previously prevailed among the UK, German, and French governments on the meaning of imminence. The UK government now diverged from the informal consensus that had existed between the three major European military-capable states and moved in its use and justification of force to a position that was convergent in practice to that espoused by the US government.

The French case is significant because while Paris has engaged in extensive uses of force in a counterterrorism role in Mali and the Sahel, it has sought to remain within the confines of conventionalist legal strategising to justify this. The French government initially restricted its acquisition and deployment of drones to an ISR role, using this enhanced capability to facilitate other means of lethal force in the counterterrorist role. The French government's use of drones changed in late 2019 when having procured an armed drone capability, it employed this means of force against terrorist targets in the Sahel. But significantly the French government has resisted the temptation to legitimate the use of this technology in terms of the expanded legal justification of self-defence proffered by London and Washington.

The case of Germany is important because it has taken an alternate position to that of the UK and French governments which has involved rejecting both the acquisition of armed drones for counterterrorism purposes and new legal interpretations that expand the boundaries of the permissible use of force. Germany is also important as the site of the most animated and politically divisive drone debate in Europe. Not only was drone acquisition and potential use a major issue for German civil society, it was also a continuous issue within various Government coalitions and between the major parties in the Bundestag. In part, this contestation was itself a consequence of Germany's constitution and the problematic nature of German debates about the use of military force for anything other than the direct defence of the Federal Republic. The German case is also important since the German government feels the need to balance its domestic sensitivities over using force outside of traditional battlefields with its core commitment to European security through its alliance with the United States in NATO.

In investigating the practice of armed drone use and associated legal strategising in the three states, we have collected data that includes primary sources such as official documents, parliamentary proceedings, and confidential semi-structured interviews with former and current practitioners as well as secondary sources (e.g. memoirs, reports, newspaper articles). While every effort has been made in a systematic and rigorous way to secure access to key officials and policymakers, political and security reasons have limited the range of interviewees prepared to talk, let alone be cited. One former UK Attorney General claimed that his time 'as AG was covered by client confidentiality' while most others stressed the sensitivity of the topic and the resultant need for secrecy. What interviews we have been able to conduct have been triangulated with other secondary sources, allowing us to develop a comprehensive understanding of how the major European military states have diverged in their use of armed drones and associated legal strategising.

Plan of the Element

The Element is divided into four sections. Section 1 shows how the UK government's legal position shifted from a rejection of an expanded definition of self-defence to a more permissive interpretation of the use of force. We argue that alongside the enabling conditions of a changing threat environment and the acquisition of new precision military technologies in the form of the drone, domestic political considerations and the leadership of David Cameron are key factors in any explanation of the changing UK position. Using the framework developed in the Introduction, we show how the UK government became a public champion in its legal strategising of a broadened conception of imminence, even though there have been no further instances to date of the UK government employing force against terrorists and invoking this legal justification.

Section 2 argues that the French government has carved out a middle path in its use of force between its desire to stay within the established principles and existing legal constraints and the need to respond to the changing threat environment at home and abroad. In navigating this middle path, the French government, as we show in the first two parts of the section, has been involved in both Africa (Mali and the wider Sahel) and the Middle East (Iraq and Syria). However, the French government has not, as yet, followed the US and UK governments in justifying the use of drones or other combat aircraft for non-battlefield targeted killing in the absence of host state consent and/or explicit/ implied UN Security Council authorisation. The final part of the section explores how the French government might seek to reconcile its fidelity to established legal rules with the use of force for counterterrorist purposes in a context where neither host state consent nor UN authorisation is available. Here, we identify three possibilities. The first is that the French government adopts an exceptionalist legal strategy to justify the use of force; the second is that it formally embraces the US/UK position, and the third is that it provides no formal legal justification at all.

The German debate over the changing parameters for the use of force has been a highly contentious one as a result of the US military response to 9/11. Section 3 argues that this complex and highly politicised debate within Germany is a key limiting factor on the possibilities for developing any public consensus at a European level on the principles that should guide the use of force. The section proceeds in three parts. First, we examine the recent protracted domestic political debate over armed drones, mapping the evolution of the debate to the formation of the 2021 coalition government and the decision for Germany to acquire armed drones for force protection purposes while maintaining a ban on Germany's military participation in non-battlefield targeted killings. Next, we show how the German government combined legal strategies of conventionalism and innovation in its decision to operate in a military support role as part of coalition air operations in Syria. The final part of the section highlights the legal tensions that Berlin has had to navigate between alliance commitments on the one hand, in the form of allowing the use of the Ramstein Air Base for US drone operations for non-battlefield targeted killing, and, on the other hand, the need to be seen to be in compliance with its own domestic and international law as interpreted by the German courts.

Having identified three case studies that exemplify different approaches to legal strategising in relation to the use of force for counterterrorism, Section 4 turns its focus to the attempts that have been made to bridge these differences at the European level. We chart the role of European Institutions, namely, the European Parliament and the European Commission in developing such a consensus. The section argues that despite pressures from these European bodies, deliberations over a new consensus have been conspicuously absent at the highest level of European Union (EU) decision-making, the European Council. We show how appeals to a common body of international law have masked divergent interpretations between European states on the permissible limits of the use of force for non-battlefield targeted killing. In making this argument, the section argues that Brexit has further complicated and problematised the search for consensus. As a consequence, discord and fragmentation continues to characterise the debate among European states and can be expected to do so for the foreseeable future.

The Conclusion revisits the puzzle motivating the Element as to what explains the divergence in legal strategising between the three European states and shows how the four key variables identified through the empirical research – domestic politics, alliance politics, individual leadership, and precedent-setting – operate in different ways in each of the cases to produce these divergences. We explore the likely contours of the changing threat environment, especially after the US withdrawal from Afghanistan in 2021 and the implications for increased uses of force as a result of the proliferation of armed drones to an ever-growing number of states. The final part of the Conclusion comes back to the relationship between international law and international politics that frames the Element; here we reflect on how far the law can and should be constraining of the use of force in contemporary international society. From our English School point of departure, we caution against the dangers of instrumentalising the law in ways that weaken the constraints on the use of force in contemporary international society more broadly.

1 The United Kingdom: 'A New Departure' in the Use of Force?

Prime Minister David Cameron announced to the House of Commons on 7 September 2015 that a Royal Air Force (RAF) Reaper drone had killed Reyaad Khan, an ISIL member of British nationality, in Raqqa, Syria, on 21 August.[10] Cameron called the event 'a new departure'. The Prime Minister explained that previously to this event, the UK's National Security Council (UKNSC) had taken the decision to authorise such military action if an opportunity presented itself. Cameron informed the House that the Attorney General, Jeremy Wright, had attended the UKNSC meeting where the decision was taken and that he had agreed that there 'was a legal basis for action'.[11] What Cameron did not say, however, was that the government's legal justification for this military action was in conformity with the non-temporal doctrine of imminence developed by the Bush and Obama administrations. We argue that two key enabling conditions explain how the UK government came to use force in August 2015. These are the juxtaposition of a changing threat environment and the acquisition of new precision military technologies in the form of the MQ-9 drone. However, what has to be explained is why it was that the Cameron government engaged in such controversial legal strategising in relation to the meaning of imminence to justify the killing of Khan. The answer given here is that the only way the Prime Minister could overcome the existing parliamentary constraints on the use of force in Syria was to invoke a broadened conception of imminence. But what is puzzling is first, why the UK government became such a strong champion after the Khan strike of such a controversial interpretation of the self-defence rule and second, why it did this without engaging in any subsequent new state practice of non-battlefield targeted killing justified in these legal terms. The section gives two answers to this puzzle.

The first is that the United Kingdom was not simply responding to US legal innovation. Rather, UK legal advisors were integral to the intellectual discussion

---

[10] Another British ISIS member, Rahul Amin, was also killed, but he was not the target of the strike.

[11] Cameron, D, HC Deb, col. 26. https://publications.parliament.uk/pa/cm201516/cmhansrd/cm150907/debtext/150907-0001.htm, 7 September 2015.

promoting the idea that the interpretation of imminence could reasonably be broadened to meet the threat from transnational terrorist groups. As a result, this new interpretation of self-defence – which had been pressed into service by the Obama administration in its use of armed drones – was readily available to UK policymakers in their legal strategising when circumstances deemed it to be necessary. Given this intellectual commitment to a broadened conception of imminence and having engaged in state practice supporting it with the killing of Khan, it was only logical that UK legal advisors would become strong advocates of this new legal rationale.

A second aspect of this puzzle is why the United Kingdom, having crossed the Rubicon of a new innovating legal defence, did not execute further military strikes of this nature. The answer offered here is that the UK strike against Khan was part of a joint operational drone campaign with the United States. As a result, even though the United Kingdom had identified high-value targets in addition to Khan, these individuals were struck by US missiles. The lack of further British strikes most likely reflected the absence of high-value targets planning attacks on the United Kingdom in the same way, or the opportunity to strike them using RAF armed drones.

The section proceeds in two parts. The first part of the section traces how the Cameron government mobilised the alternative broad interpretation of imminence to justify the UK's killing of Khan and the second part focuses on the UK government's participation in the allied coalition air operations over Syria, showing how its legal justifications embody a mix of conventional and innovating claims.

The United Kingdom Adopts a Doctrine of Non-Temporal Imminence

In announcing to parliament that a RAF drone strike had killed Khan in August 2015, Cameron said, 'this [is] the first time in modern times that a British asset has been used to conduct a strike in a country where we are not involved in a war'. He explained that the UK government was:

> Exercising the UK's inherent right to self-defence. There was clear evidence of these individuals planning and directing armed attacks against the UK. These were part of a series of actual and foiled attempts to attack the UK and our allies. And in the prevailing circumstances in Syria, the airstrike was the only feasible means of effectively disrupting the attacks that had been planned and directed.[12]

---

[12] Cameron, D., Syria: refugees and counter-terrorism – Prime Minister's statement, www .gov.uk/government/speeches/syria-refugees-and-counter-terrorism-prime-ministers-statement, 7 September 2015.

In order to understand Cameron's specific legal justification of 'individual self-defence', it is necessary to appreciate the constraints that parliament had imposed on the government using force in Syria. As we will discuss later in the section, the House of Commons voted on 26 September 2014 to adopt a resolution that provided for the UK military to use force against ISIL in response to the request for assistance from the Government of Iraq in support of its collective self-defence. Specifically, however, the text of the resolution 'notes that this motion does not endorse UK air strikes in Syria as part of this campaign and any proposal to do so would be subject to a separate vote in Parliament'.[13]

Given this clear constraint imposed by parliament, Cameron was faced with a choice between either seeking new authority from the legislature to target ISIS members in Syria or have recourse to the argument that there was an imminent threat to the United Kingdom that could only be met through the immediate use of lethal action (Birdsall, 2022: 208). Considerations of secrecy and the perceived immediacy of the threat precluded the former and for Cameron, and in the view of the National Security Council, the only choice was to use force.

The constitutionality of this decision not to go back to parliament was subsequently supported by the parliamentary Joint Committee on Human Rights (JCHR) in its 2016 report into the 'The Government's Policy on the Use of Drones for Targeted Killing'. The Committee recognised that while a 'constitutional convention' has emerged that requires parliament to approve the use of force, this can be superseded 'when there is an emergency which means it would not be appropriate to consult the House of Commons in advance. Examples of such exceptions include if there were a critical British national interest at stake, or considerations of secrecy make it impossible'.[14]

In setting out his defence of the military action the Prime Minister did not explicitly explain how these individuals presented an 'imminent' threat to UK citizens. Critics of this new policy were quick to highlight this lack of specific evidence. For example, David Davis MP, chairing an emergency meeting of the All-Party Parliamentary Group on Drones (APPG) on 16 September said, 'As far as I'm aware there have been no arrests in relation to this in the UK . . .. That does rather drive a coach and horses through that argument'.[15] At the same

---

[13] Cameron, D., 'Iraq: Coalition Against ISIL', 26 September 2014, https://hansard.parliament.uk/commons/2014-09-26/debates/1409266000001/IraqCoalitionAgainstISIL.

[14] Joint Committee on Human Rights, 'The Government's Policy on the Use of Drones for Targeted Killing', https://publications.parliament.uk/pa/jt201516/jtselect/jtrights/574/574 .pdf, 2016, p. 45.

[15] Ross, A, 'Tory MP casts doubt on justification for drone strikes on Britons in Syria'. *The Guardian,* 16 September 2015, www.theguardian.com/world/2015/sep/16/drone-strikes-brit ons-syria-tory-mp-casts-doubt-justification.

meeting, former Director of Public Prosecutions, Sir Keir Starmer (then a backbench MP), expressed his concern at what he saw as the Attorney General's reinterpretation of the principle of imminence. A year after the drone strike against Khan, the JCHR concluded in its report that the UK government had moved to adopt 'a more flexible approach to the meaning of "imminence", to include an ongoing threat of a terrorist attack from an identified individual who has both the intent and the capability to carry out such an attack without notice'.[16]

In defending the decision to kill Khan, Secretary of State for Defence, Michael Fallon, stated: 'We wouldn't hesitate to do it again if we knew there was an armed attack planned and we knew who was behind it'.[17] Given this philosophy on the part of UK policymakers, what is puzzling is why the United Kingdom did not engage in subsequent strikes of this nature. An explanation for this can be deduced from Cameron's memoirs. He recalled that in May 2015 he had convened a 'Ministerial Small Joint Group meeting that tasked the relevant authorities to locate and kill – attack planners. Several targets were identified' (Cameron, 2019: 598). These included three British nationals Khan, Junaid Hussain, and Mohammed Emwazi ('Jihadi John'). Despite Hussain and Emwazi being identified as British high priority targets, both were killed in US airstrikes in August and November 2015. The RAF's killing of Khan in August 2015 was part of a wider counterterrorism air operation between the two governments against ISIL operatives in Syria. Cameron was clearly willing and indeed expecting to take further strikes, but as it turned out, these missions were conducted by US drones. As Cameron expressed it, 'we and the Americans were working hand in glove, sharing the burden, and both ready to strike if the conditions were right' (2019: 598). During these drone operations, both US and UK Reaper aircraft would work as a team, providing live video coverage and targeting options. Which aircraft ultimately took the strike was less dependent on the national ownership of the drone, than on the position of the individual assets at the best moment for the strike.[18] Part of the explanation for this was also that the RAF had a much more limited supply of armed drones compared to the Americans with the result, according to a well-placed source, that the United Kingdom never seemed to have its military capabilities in the right place at the right time.[19]

---

[16] 'The Government's policy on the use of drones for targeted killing', House of Lords, House of Commons Joint Committee, publications.parliament.uk/pa/jt201516/jtselect/jtrights/574/574.pdf, 2016, p. 47.

[17] Quoted in Jenny Booth, 'We will not hesitate to kill more Britons, says Fallon', *The Times*, 8 September 2015, www.thetimes.co.uk/article/we-will-not-hesitate-to-kill-more-britons-says-fallon-237rnb8k3r3.

[18] Confidential interview with former senior US military official, 18 May 2017.

[19] Confidential interview with former senior UK official (A), 4 January 2023.

The other puzzle that we have identified in relation to the British case concerns how it was that having opposed a broad interpretation of imminence in the early 2000s, most notably in relation to the 2003 Iraq War,[20] the UK government became such a strong advocate of it after the Khan strike. Although there were strong domestic political reasons motivating the Prime Minister to use the imminence argument, the UK government showed that it was not only an act of political convenience by employing this controversial legal rationale in its Article 51 letter defending the action to the UN Security Council. Echoing Cameron's defence the previous day in the House of Commons, Matthew Rycroft, the UK's Permanent Representative to the UN, invoked the expanded imminence argument to justify a 'precision airstrike against an ISIL vehicle in which a target known to be actively engaged in planning and directing imminent armed attacks against the United Kingdom was travelling'.[21] However, the UK government did not only rely on this legal argument and in fact the first legal justification used in Rycroft's letter was the more established conventionalist legal rationale of the action being in support of the collective self-defence of Iraq. Rycroft justified the Khan strike as being in conformity with the government's earlier justification of the RAF's participation using both drones (particularly in an Intelligence, Surveillance, and Reconnaissance role) and combat aircraft in the US-led coalition air operations against ISIL in Iraq.[22] He wrote, 'ISIL is engaged in an ongoing armed attack against Iraq, and therefore action against ISIL in Syria is lawful in the collective self-defence of Iraq'.[23]

---

[20] The then Attorney General, Lord Peter Goldsmith, challenged what he viewed as the US reinterpretation of imminence to justify the use of force to disarm Iraq of WMD. In a minute on 7 March 2003 setting out his formal legal advice that was sent only to Prime Minister Tony Blair (though was seen by other ministers), the Attorney General clearly understood imminence as a temporal concept. He wrote, 'I am aware that the USA has been arguing for recognition of a broad doctrine of a right to use force to pre-empt danger in the future. If this means more than a right to respond proportionately to an imminent attack (and I understand that the doctrine is intended to carry that connotation) this is not a doctrine which, in my opinion, exists or is recognised in international law' (The full minute was published in *The Guardian* in April 2005. See 'Full text: Iraq legal advice', 28 April 2005,www.theguardian.com/politics/2005/apr/28/election2005.uk (for further discussion of the UK's legal debate, see Sands, 2005; Greenstock, 2016).

[21] Letter to the President of the Security Council from the UK's Permanent Representative to the UN, 8 September 2015, UN Doc S/2015/688, www.securitycouncilreport.org/atf/cf/% 7B65BFCF9B-6D27-4E9C-8CD3-CF6E4FF96FF9%7D/s_2015_688.pdf.

[22] The UK's Permanent Representative to the UN had submitted a letter to the President of the Security Council on 26 November 2014 in which Ambassador Mark Lyall Grant explained that the UK would be 'taking measures in support of the collective self-defence of Iraq as part of international efforts led by the United States' (Letter to the President of the Security Council from the UK's Permanent Representative to the United Nations, UN Doc S/2014/851, 26 November 2014, www.securitycoun cilreport.org/atf/cf/%7B65BFCF9B-6D27-4E9C-8CD3-CF6E4FF96FF9%7D/s_2014_851.pdf.

[23] Letter to the President of the Security Council from the UK's Permanent Representative to the UN, 8 September 2015, UN Doc S/2015/688, www.securitycouncilreport.org/atf/cf/% 7B65BFCF9B-6D27-4E9C-8CD3-CF6E4FF96FF9%7D/s_2015_688.pdf. See also Gray 2016, 199.

One interpretation of this legal strategising is that the Cameron government wanted to first and foremost ground the action in the conventionalist legal justification of self-defence. However, ministers and legal advisors also wanted to make the case that absent the justification of 'collective self-defence', there would still be clear legal grounds for the Khan strike under the broadened interpretation of imminence.

Several explanations are possible for the different emphasis between Cameron's statement and Rycroft's letter to the UN. One former senior official suggested that the letter represented a 'belt and braces' approach, 'throwing everything you have at it to make the case',[24] while a further explanation could be that perhaps there was a divergence in thinking between Number 10 and the foreign office lawyers who were keen to locate Cameron's innovative legal strategising within a well-established conventionalist legal framework.[25]

The UK government's commitment to defending and promoting a broadened conception of imminence was left in no doubt however, and it was clearly demonstrated when the UK's Attorney General delivered a keynote address at the International Institute for Strategic Studies (IISS) in London. Wright defended the broadened conception of imminence as a justifiable and appropriate response to the threat posed by non-state terrorist groups operating transnationally. His speech, and the wider innovating legalist strategy it belonged to,[26] illustrates the Skinnerian manoeuvre of referring to existing principles and rules, while at the same time reinterpreting them in new contexts to legitimate actions that would previously have been viewed as illegitimate under these principles. To this end, Wright argued that, 'International law is not static and is capable of adapting to modern developments and new realities .... The phenomenon of international terrorism, for example, has caused the international community to apply the law to new circumstances'.[27] Specifically in relation to imminence, Wright used his speech to argue that what 'we ... need to think

---

[24] Confidential interview with former senior UK official (B), 2 November 2022.
[25] Confidential interview with former senior UK official (A), 4 January 2023.
[26] Cameron's successor, Theresa May, submitted in December 2017 a written statement to the Intelligence and Security Committee after the Committee had reported in April 2017 on the intelligence aspects of the Khan strike (Intelligence and Security Committee of Parliament report on *UK Lethal Drone Strikes in Syria*, 26 April 2017, https://isc.independent.gov.uk/wp-content/uploads/2021/01/20170426_UK_Lethal_Drone_Strikes_in_Syria_Report.pdf). May said in her letter that, 'A direct and imminent threat [from Khan and Amin] was identified by the intelligence agencies and the National Security Council agreed that military action should be taken' (May, 'Intelligence Oversight', Written statement HCWS378, 20 December 2017, www.parliament.uk/business/publications/written-questions-answers-statements/written-statement/Commons/2017-12-20/HCWS378/).
[27] Quoted in Wright, J., *'Attorney General's speech at the International Institute for Strategic Studies'*, 11 January 2017, London www.gov.uk/government/speeches/attorney-generals-speech-at-the-international-institute-for-strategic-studies.

about as a society … is what imminence means in the context of a terrorist threat, compared with back in the 1830s when the customary international law test was set down following the Caroline Incident. When do we now say a threat of an armed attack is sufficiently imminent to trigger a state's right to use force in self-defence?'

The change in emphasis partly reflected the Cameron government's frustration between the intermittent intelligence picture and the legal requirement for the threat to be immediately imminent in order to justify an attack. Broadening the government's conception of imminence was seen as a means of aligning the intelligence on planned terrorist actions, the capability to act, and the right to do so in a way that allowed a military response.[28]

In setting out the UK government's position on imminence, Wright acknowledged the importance of the intellectual contribution of Sir Daniel Bethlehem, the FCO's Principal Legal Advisor between May 2006 and May 2011. Shortly after leaving office, he set out in an article published in 2012 the principles 'relevant to the actual circumstances in which states are faced with an imminent or actual armed attack by nonstate actors'. What is most important here is 'Principle 8' where Bethlehem sets out the criteria that must be met for a threat to be assessed as imminent. He wrote:

> Whether an armed attack may be regarded as 'imminent' will fall to be assessed by reference to all relevant circumstances, including (a) the nature and immediacy of the threat, (b) the probability of an attack, (c) whether the anticipated attack is part of a concerted pattern of continuing armed activity, (d) the likely scale of the attack and the injury, loss, or damage likely to result therefrom in the absence of mitigating action, and (e) the likelihood that there will be other opportunities to undertake effective action in self-defense that may be expected to cause less serious collateral injury, loss, or damage (Bethlehem, 2012: 6).

The key implication that followed from applying these criteria was that imminence could be framed in non-temporal terms and this led Bethlehem to claim that: 'The absence of specific evidence of where an attack will take place or of the precise nature of an attack does not preclude a conclusion that an armed attack is imminent for purposes of the exercise of a right of self-defense, provided that there is a reasonable and objective basis for concluding that an armed attack is imminent' (Bethlehem, 2012: 6–7).

While Bethlehem was careful to accept full responsibility for the ideas in his article, he also acknowledged that they were 'informed by detailed discussions over recent years with foreign ministry, defense ministry, and military legal

---

[28] Confidential interview with former senior UK official (A), 4 January 2023.

advisers from a number of states who have operational experience in these matters' (Bethlehem, 2012: 4). Bethlehem's role as the FCO's Principal Legal Advisor would have meant he was prominently involved in these discussions. The reference to a 'number of states who have operational experience in these matters' clearly refers to the United States and United Kingdom and strongly suggests that Bethlehem and other UK legal advisors played a prominent role in the co-production of the broadened conception of imminence. For Bethlehem, the process by which 'the principles' had been jointly arrived at was an important vehicle in forging a wider international consensus on 'the contours of the law relevant to the actual circumstances in which states are faced with an imminent or actual armed attack by nonstate actors'. Wright indicated in his lecture that in the first instance this consensus was being forged within the Five Eyes grouping comprising the United States, Canada, Australia, New Zealand, and the United Kingdom. Here, he made specific reference to a 2016 meeting of the Quintet of Attorney-Generals from these five states, though he did not elaborate further how far there was a legal consensus within the grouping on the broadened interpretation of imminence.[29]

UK legal advisors, then, sought through the vehicle of Wright's speech to position the UK government as a leading advocate for a broadened conception of imminence, while claiming at the same time that they were not legally innovating, but only applying existing principles of international law. Such an innovating legal strategy failed to satisfy many international lawyers, human rights NGOs, and parliamentarians who were quick to contest it.

Two important parliamentary investigations in 2016 and 2018 respectively challenged Wright and the government's claim that a broadened conception of imminence was both legal and in the long-term security interest of the United Kingdom. The JCHR in its 2016 report raised 'concerns about the implications of too expansive a definition of "imminence" for the width of the right of self-defence in international law'.[30] The Committee asked in its report, in the

---

[29] Australian Attorney General, George Brandis, in May 2017 delivered a speech that was almost a carbon copy of Wright's IISS speech (according to one well-placed source, the speech was 'simply handed over by the UK'). Echoing Wright on the Caroline case, Brandis said, 'technology and the nature of national security threats have been revolutionised' so much so that 'they would be utterly unrecognisable to Secretary Webster' (Brandis, G., 'The Right of Self-Defence Against Imminent Armed Attack In International Law', www.ejiltalk.org/the-right-of-self-defence-against-imminent-armed-attack-in-international-law/#more-15255.

[30] 'The Government's policy on the use of drones for targeted killing', House of Lords, House of Commons Joint Committee, https://publications.parliament.uk/pa/jt201516/jtselect/jtrights/574/574.pdf, 2016, p. 47. Yasmine Ahmed, Executive Director of Rights Watch UK, warned that: 'The UK's changed position on targeted killings represents a radical departure from established international law. It risks unpicking the fabric of the collective security system crafted after the Second World War' (2018).

context of the Khan strike, whether 'evidence that an individual is planning terrorist attacks in the UK' is enough to trigger a right of self-defence? Or, 'does the preparation need to have gone beyond mere planning?' And once 'a specific individual has been identified as being involved in planning or directing attacks in the UK', is the threat now 'in effect, permanently imminent?'[31] This concern about the enabling possibilities for the use of force entailed by a reinterpretation of imminence was expressed two years later by the APPG on Drones in its report on the UK's use of armed drones.

The APPG report, produced after a year-long enquiry, and chaired by Professor Michael Clark with Professor Dapo Akande as its legal advisor, identified the linkage between the principles set out by Bethlehem in his 2012 article and the UK's formal and fully articulated endorsement of these principles in Wright's 2017 speech. The APPG report questioned the legal standing of the government's reliance on the 'Bethlehem Principles'. The Committee wrote:

> In the view of some commentators, Sir Daniel's Principles represent a radical expansion of the right of anticipatory self-defence by providing a new standard of imminence to enable preemptive military strikes against threats ... the UK's actions could have knock-on effects for the stability of the international rules-based order. The long-term implications of an expansive definition of 'imminence' are the potential erosion of use of force norms more broadly in a manner that may be used by an increasingly greater number of states, including states such as Russia, North Korea and China.[32]

The APPG enquiry was drawing attention in this passage to the legally contested nature of the US and UK government's reinterpretation of imminence. The Committee's contention that the 'Bethlehem principles' are not an 'authoritative statement on the law in this area' is supported by the lack of *opinio juris* from other states (with the exception of the US and Australian governments) supporting the UK's broadened conception of imminence. In addition, and reinforcing the JCHR's warning in its 2016 report, the APPG emphasised in this passage the deleterious consequences of further eroding the general ban on the use of force in Article 2 (4) of the UN Charter through what it called 'a radical expansion of the right of anticipatory self-defence'.[33]

---

[31] 'The Government's policy on the use of drones for targeted killing', House of Lords, House of Commons Joint Committee, https://publications.parliament.uk/pa/jt201516/jtselect/jtrights/574/574.pdf, 2016, p. 47.

[32] APPG on Drones, *The UK's Use of Armed Drones: Working With Partners*, https://appgdrones.org.uk/wp-content/uploads/2014/08/INH_PG_Drones_AllInOne_v25.pdf, July 2018, p. 37 (see also Byers, 2003: 182; Ahmed, 2018).

[33] APPG on Drones, *The UK's Use of Armed Drones: Working With Partners*, p. 37, https://appgdrones.org.uk/wp-content/uploads/2014/08/INH_PG_Drones_AllInOne_v25.pdf, July 2018.

As we discuss in the next section, UK airpower was employed in Syria after 2015 but the Cameron government did not need to rely on the doctrine of imminence to justify this because parliamentary authority was now forthcoming to support this, obviating the need to rely on the controversial justification of imminence that Cameron had used to defend the killing of Khan to parliament.

Justifying Coalition Air Strikes over Syria: The Cameron Government Combines Conventionalist and Innovating Legal Strategies

On 29 August 2013, Cameron sought parliamentary approval to join the United States in undertaking military action against Syria for its use of chemical weapons. Opening the debate, he said that such action would be 'legal, proportionate and focused on saving lives by preventing and deterring further use of Syria's chemical weapons'.[34] The Prime Minister told the House of Commons that 'the use of chemical weapons is a war crime … a crime against humanity [and] the principle of humanitarian intervention provides a sound legal basis for taking action'.[35] On the day of the parliamentary vote, the government published a summary of the advice of the Attorney General, Dominic Grieve, that set out the UK's doctrine of humanitarian intervention. Given the precarious nature of this legal principle to justify the use of force (Franck and Rodley, 1973; Wheeler, 2000), Grieve was careful to argue that the UK government was seeking, with its French and US allies, a new UN resolution under Chapter VII that would provide authority to member states to 'take all necessary measures to protect civilians in Syria from the use of chemical weapons and prevent any future use of Syria's stockpile of chemical weapons'. But such a resolution had not been secured by the time the vote was taken.

However, the Attorney General emphasised in his note that, 'If action in the Security Council is blocked, the UK would still be permitted under international law to take exceptional measures in order to alleviate the scale of the overwhelming humanitarian catastrophe in Syria by deterring and disrupting the further use of chemical weapons by the Syrian regime'.[36] After debating the Government's motion to use force, parliament voted against it by 285 votes to 272. A key factor in the defeat of the government's motion was the lack of a new UN resolution, since many Labour MPs, including the Leader of the Opposition

---

[34] Cameron, D., HC Deb, 29 August 2013, col. 1425. https://publications.parliament.uk/pa/cm201314/cmhansrd/cm130829/debtext/130829-0001.htm.

[35] Cameron, D., HC D, (29 August 2013). Column. 1426. https://publications.parliament.uk/pa/cm201314/cmhansrd/cm130829/debtext/130829-0001.htm.

[36] Chemical weapon use by Syrian regime: UK government legal position, 29 August 2013, www.gov.uk/government/publications/chemical-weapon-use-by-syrian-regime-uk-government-legal-position/chemical-weapon-use-by-syrian-regime-uk-government-legal-position-html-version.

Ed Miliband, argued that they would have voted differently had the action had the seal of UN approval. Miliband stated that 'People are deeply concerned about the chemical weapons attacks in Syria, but they want us to learn the lessons of Iraq . . .. They don't want a rush to war. They want things done in the right way, working with the international community'.[37]

The UK government's involvement in the Syrian conflict took its next turn in September 2014. The Iraqi government, suffering military losses from ISIL (which was becoming a key military and political presence in the region), wrote to the President of the Security Council requesting military assistance.[38] This request grew out of the US government and twenty-six other states committing on 15 September at a conference in Paris to come to the aid of Iraq by engaging ISIL's forces inside Iraq. Following these developments, the UK government invited parliament to approve its plans to assist Baghdad 'in protecting civilians and restoring its territorial integrity, including the use of UK air strikes to support Iraqi, including Kurdish, security forces' efforts against ISIS in Iraq'.[39] Cameron made clear to parliament that this vote would not endorse military strikes in Syria, only in Iraq.[40]

The legal basis for this proposed use of force was the conventionalist one that the Iraqi government had exercised its sovereign right to call for military assistance and in responding to this, the United Kingdom was acting in the collective self-defence of Iraq.[41] The UK government's innovating legalist strategy to justify the use of force in Syria in the form of a doctrine of unilateral humanitarian intervention had failed to muster parliamentary approval in August 2013. But in September 2014, the House of Commons overwhelmingly supported the Prime Minister's legal conventionalist strategy to use force to defend Iraq with the motion carried by 524 votes to 43.

The RAF, using GR4 Tornadoes, carried out its first strike in support of Kurdish forces in Northwest Iraq fighting against ISIL terrorists on 30 September 2014. The first UK drone strike against ISIL forces in Iraq was

---

[37] Miliband, E., 'Syria and the use of Chemical Weapons', 29 August 2013, https://hansard.parliament.uk/commons/2013-08-29/debates/1308298000001/SyriaAndTheUseOfChemical Weapons.

[38] Letter to the President of the Security Council from Iraq's Permanent Representative to the UN, UN Doc S/2014/691, 8 September, 2014, www.un.org/en/ga/search/view_doc.asp?symbol=S/2014/691.

[39] House of Commons Debate (26 September 2014). Volume. 585. Column. 1256. https://hansard.parliament.uk/Commons/2014-09-26/debates/1409266000001/IraqCoalitionAgainstISIL.

[40] Cameron, D., 'Iraq: Coalition Against ISIL (opening statement)', 26 September 2014, https://publications.parliament.uk/pa/cm201415/cmhansrd/cm140926/debtext/140926-0001.htm.

[41] Cameron, D., HC Deb, Volume. 585. Column. 1263, 26 September 2014, https://hansard.parliament.uk/Commons/2014-09-26/debates/1409266000001/IraqCoalitionAgainstISIL.

on 10 November with a RAF Reaper firing a Hellfire missile at ISIL forces in Baiji, 140 miles north of Baghdad.[42] But while the UK military's legal authority to attack ISIL forces was confined to Iraqi territory given the Prime Minister's commitment to parliament during the debate on 26 September, RAF combat aircraft and drones were carrying out ISR missions on Syrian territory from late 2014, including providing targeting intelligence for strikes by allied militaries participating in the US-led coalition (Holland, 2020: 181).

ISIL's military successes in 2015 led it to gain an increasing hold over territory in Syria and its growing military assertiveness was accompanied by a series of terrorist atrocities outside Syria perpetrated against what they regarded as the enemies of their brand of Sunni Islam. Thirty British citizens (and three others) were killed on 26 June 2015 in Tunis by ISIL terrorists in a mass shooting. Following this atrocity, UK Defence Secretary Fallon claimed that ISIL needed to be attacked 'at source', meaning Syria, though he accepted at this time that this was not possible unless there was a parliamentary vote supporting this.[43] In the aftermath of the Tunis attack, UK public opinion was largely sympathetic to the rationale behind the strikes and further terrorist acts by ISIL reinforced the public and elite perceptions of threat. ISIL struck again, killing innocent civilians in Ankara on 10 October and again in Beirut on 12 November. The deaths of 130 civilians in a series of coordinated terror attacks in Paris on 13 November 2015 further deepened the sense of threat among UK policymakers.

It was in this general context of rising threat perception, particularly the Tunis attacks, that the UK government in September 2015 publicly justified the killing of Khan (that had taken place the previous month) as an act of self-defence. As we have discussed, using an expanded definition of imminence to justify this lethal military action had generated both legal and political controversy. To try and secure increased domestic legitimacy for future strikes of this kind, Cameron sought new legal authority to extend the RAF's attacks against ISIL in Iraq to its safe havens in Syria. Four days after the Paris attacks, he said in parliament:

> There is no Government in Syria with whom we can work, particularly in that part of Syria. There are no rigorous police investigations or independent courts upholding justice in Raqqa. We have no military on the ground to detain those preparing plots against our people . . .. We face a direct and

---

[42] Norton-Taylor, R., 'UK launches first drone strike in Iraq against Isis militants', *The Guardian*, 10 November 2014, www.theguardian.com/world/2014/nov/10/uk-first-drone-strike-iraq-isis.

[43] Fallon was quoted in a BBC News article. See, 'Consider Syria IS strikes, defence secretary urges MPs', 3 July 2015, www.bbc.co.uk/news/uk-33358267.

growing threat to our country, and we need to deal with it not just in Iraq but in Syria too . . . and the case for doing it has only grown stronger after the Paris attacks.[44]

What is noteworthy about Cameron's justification for expanding UK strikes to Syria in this speech is his invocation of the unable and unwilling justification in the context of the Syrian government's failure to prevent ISIL from operating on Syrian territory. The FCO's legal advisors argued that there was a secure legal basis in Article 51 of the UN Charter for the UK government to use force against a non-state terrorist group like ISIL, where the government of the state whose territory force is being used on is 'unwilling and/or unable to take action necessary to prevent ISIL's continuing attack on Iraq, or indeed attacks on us'.[45]

The UK government's use of the unable or unwilling rationale as a key part of its self-defence claim is an example of innovating legalist strategising at work, given that the doctrine did not have a strong grounding in customary inter-national law (Brunee and Toope, 2011). However, the UK government was not comfortable basing its legal defence of the use of force in Syria on this argument alone. Instead, it sought a stronger conventionalist legal justification in the form of a new UN Security Council resolution that would complement its Article 51 claim. With British diplomats playing a key supporting role, this resolution was adopted on 20 November 2015 as Resolution 2249, which called on member states 'to eradicate the safe haven they [ISIL] have established over significant parts of Iraq and Syria'.[46]

The Resolution was not passed under Chapter VII of the UN Charter which would have provided the most solid legal basis for the use of force (Akande and Milanovic, 2015). Nevertheless, it was (deliberately) sufficiently ambiguous in its wording (Byers, 2021: 104–7) that the UK prime minister, in proposing a motion on 2 December to use force in Syria, felt emboldened to claim alongside the unable and unwilling justification that the United Kingdom was on secure legal ground in using force against ISIS in Syria.[47] Cameron reminded the House that

---

[44] Cameron, D., 'Prime Minister's statement on Paris attacks and G20 Summit', 17 November 2015, www.gov.uk/government/speeches/prime-ministers-statement-on-paris-attacks-and-g20-summit.

[45] This legal reasoning was set out in Cameron's Memorandum to the Foreign Affairs Select Committee ('Memorandum to the Foreign Affairs Select Committee. Prime Minister's Response to the Foreign Affairs Select Committee's Second Report of Session 2015–16: The Extension of Offensive British Military Operations to Syria', 14 November 2015, www.parliament.uk/docu ments/commons-committees/foreign-affairs/PM-Response-to-FAC-Report-Extension-of-Offensive-British-Military-Operations-to-Syria.pdf).

[46] UN Security Council Resolution 2249, UN Doc S/Res/2249 (2015), 20 November 2015, www .securitycouncilreport.org/atf/cf/%7B65BFCF9B-6D27-4E9C-8CD3-CF6E4FF96FF9%7D/ s_res_2249.pdf.

[47] Cameron D., House of Commons Debate Col. 323. (2 December 2015, https://publications .parliament.uk/pa/cm201516/cmhansrd/cm151202/debtext/151202-0001.htm.

'ISIL has brutally murdered British hostages. They have inspired the worst terrorist attack against British people since 7/7 on the beaches of Tunisia, and they have plotted atrocities on the streets here at home'.[48] He asked his fellow MPs rhetorically: 'Do we work with our allies to degrade and destroy this threat, and do we go after these terrorists in their heartlands, from where they are plotting to kill British people, or do we sit back and wait for them to attack us?' His own answer was clear: 'We possess the capabilities to reduce this threat to our security, and my argument today is that we should not wait any longer before doing so. We should answer the call from our allies'.[49] The government won this vote 397 to 223, and in his Article 51 letter to the UN Security Council the following day, the UK Permanent Representative to the UN cited both Resolution 2249 and the 'inherent right of individual and collective self-defence' as legal justifications.[50] In doing so, the UK government combined *conventional* and *innovating* legalist strategies to justify its extension of force to Syria, demonstrating the pragmatic strategising by which the UK government tailored its legal justifications to meet different exigencies for the use of force.

Conclusion

Two puzzles motivate this section: first, why did the Cameron government adopt a form of legal strategising in relation to self-defence against non-state actors that was such a radical departure from the legal justifications employed by previous UK governments? And second, why did it become such a champion of this? Turning to the first question, Cameron needed to make the imminence argument in order to circumvent the constraints imposed on his government by the parliamentary resolution limiting the UK's use of force to Iraq and explicitly prohibiting its extension to Syria. While Cameron's hand was forced domestically by the legal constraints, he was mindful of the need to ensure that decision-making on the use of force rested on firm legal grounds.[51] As he explained in his memoir, 'I wanted to know what was and what was not legal as we formulated policy at its earliest stages, not just at the end of the process' (Cameron, 2019: 599). This shows Cameron's recognition of how law can be a constraining

---

[48] Cameron, D., House of Commons Debate Col. 325, 2 December 2015, https://publications.parliament.uk/pa/cm201516/cmhansrd/cm151202/debtext/151202-0001.htm.

[49] Cameron, D., House of Commons Debate Col. 325, 2 December 2015, https://publications.parliament.uk/pa/cm201516/cmhansrd/cm151202/debtext/151202-0001.htm.

[50] Letter to the President of the Security Council from the Permanent Representative of the United Kingdom of Great Britain and Northern Ireland, UN Doc S/2015/928, 3 December 2015, https://securitycouncilreport.org/atf/cf/%7B65BFCF9B-6D27-4E9C-8CD3-CF6E4FF96FF9%7D/s_2015_928.pdf.

[51] One salient earlier manifestation of this was his decision to appoint the first legal advisor to the Prime Minister's Office in 2014.

influence on policymaking, but equally relevant as far as he was concerned, it was important to explore how far there could be permissive innovations within the law to enable new policy actions that previously might have been ruled out on legal grounds. He wrote in his memoir that it was important to consider 'what might work and then ask lawyers – not the other way round' (Cameron, 2019: 599).

In terms of the second puzzle as to why the Cameron government became a champion of a broadened conception of imminence after the killing of Khan in August 2015, this was not just convenient legal strategising confined to the circumstances of the moment. Instead, a non-temporal conception of imminence had grown out of the intellectual gestation between UK legal officials and their US and other allied counterparts in the preceding years (Bethlehem, 2012: 4). Before August 2015, the UK government had not put into practice a broadened conception of imminence. This perhaps, in part, reflected the views of David Cameron's first Attorney General, Dominic Grieve, in post from May 2010 until July 2014, who according to one source held a more restrictive view of what international law permitted with regard to the use of force in self-defence.[52] His successor, Jeremy Wright, in post from July 2014, according to this source held a more permissive interpretation of the criterion of imminence.[53] Just under a year after this change of Attorney General, the UK government engaged in its first non-battlefield targeted killing and justified this in terms of the broadened doctrine of imminence. The change of Attorney General in the context of an increased threat environment ensured that when Cameron sought a legal opinion on the killing of Kahn and the other two UK nationals identified by senior ministers in mid-May 2015, the response was a favourable one.

Faced with considerable domestic criticism after the strike, the government responded, not by retreating back to a position of legal conventionalism, but instead strongly advocating for a broadened conception of imminence. In seeking to change the law in this way, despite claiming to be in conformity with it, ministers and their legal advisors argued that international law had to be adapted to meet the exigencies of the threat from transnationally based non-state terror groups in a post 9/11 world.

The related legal innovation of the Cameron premiership was the extension of the right to use force in self-defence against non-state terrorist groups through the use of the unable and unwilling argument. The claim being that Syria, as a result of being an ungoverned space, was unable to prevent ISIL using its

---

[52] Confidential interview with former UK senior official (A), 4 January 2023.
[53] Confidential interview with former UK senior official (A), 4 January 2023.

territory to attack Iraq. However, the Cameron government appreciated that securing new parliamentary authority for the use of force against ISIL in Syria would be strongly facilitated by a new UN Security Council resolution supporting this. In seeking conventionalist legal authority to strike targets in Syria, the government came close to admitting that the unable and unwilling justification lacked a strong grounding in customary international law and that as a result, it would be less likely to secure domestic legitimation, especially among opposition MPs, compared to the recognised legal authority that comes from a Chapter VII Security Council resolution. Despite the ambiguities as to just what Resolution 2249 provided in terms of legal authority for the use of force, it was sufficient to secure the parliamentary approval that placed UK strikes in Syria on much firmer legal ground than that provided by the unable and unwilling doctrine.

What motivated Cameron to position his government in a legal innovating role was the pressure of alliance commitments and his own moral compass in responding to the threat posed by violent non-state actors. With regard to the former, Obama told Cameron on a trip to Washington in January 2015 that, 'I totally understand your position [i.e., Cameron's lack of a parliamentary mandate to use force in Iraq and Syria] but I'd really value more action if you win the election' (quoted in Seldon and Snowdon, 2016: 465–6). Cameron was eager to respond to Obama's request and show solidarity with the US and French governments over Syria. This was the context for his rhetorical appeal that: 'Do we work with our allies to degrade and destroy this threat'. His answer was unequivocally that, 'We should answer the call from our allies'[54] (see also Cameron, 2019: 604–5).

Cameron's personal determination that his government play a leading role in the fight against ISIL stemmed, in part, from his own personal convictions about the scale of the threat facing the United Kingdom. According to Anthony Seldon and Peter Snowdon writing in 2016, 'The threat of a terrorist attack on Britain continues to be one topic that keeps Cameron awake at night' (2016: 542). Cameron reflected in his 2019 memoir that he felt there was not enough willingness on the part of the national security bureaucracy to come up with new ideas and approaches to meet what he perceived as a compelling mortal challenge to the UK's security. For the Prime Minister: 'Our military and security services were, on this issue, a huge source of frustration ... "Maybe it's just a wicked problem that cannot be solved", was the gist. I made the point that just because we didn't have the ability to achieve change on our own, that

---

[54] Cameron, D., House of Commons Debate Col. 325, (2 December 2015), https://publications .parliament.uk/pa/cm201516/cmhansrd/cm151202/debtext/151202-0001.htm.

wasn't a decisive argument for holding back altogether'. Later in his memoir he wrote of his growing frustrations: 'Why, I kept asking weren't we doing more ourselves, particularly now that the ISIS threat was growing? Why should we accept and work with the US programme, but not supplement it with our own?' (Cameron, 2019: quotes at 452, 599). For Cameron, the United Kingdom had a moral responsibility to shoulder its share of the burden in the battle against ISIL, and as prime minister, he was determined to see this happen.

A key enabling condition of the UK government playing the role of a responsible ally was that the UK military had, in the MQ-9 Reaper, a capability that could make a potent contribution to the collective allied counterterrorism endeavour. As Cameron wrote in his memoir, 'We had the necessary drones, and were getting more. We were key to the intelligence, including on the ground, that could help target those who intended to do us harm' (Cameron, 2019: 599). The Prime Minister made this a central component of his speech to the House of Commons, stating that, 'We possess the capabilities to reduce this threat to our security, and my argument today is that we should not wait any longer before doing so'.[55] In a sense this was an important corollary of the unwilling and unable legal justification that the United Kingdom had employed justifying its use of force in Syria. Now, largely due to its armed drone capability, the UK government was itself 'able and willing' to conduct these operations. That is to say, it was not just the threat and the legal justification that was innovative, the ability to identify, track and strike targets at reach with impunity was an important part of these new developments. What these new technological capabilities made possible was a joint mode of operation with US political and military authorities which, as Cameron stated in his memoirs, ensured that 'we and the Americans were working hand in glove' (Cameron, 2019: 598). What this meant in practice was that the UK government did not necessarily need to strike all the individuals that it had identified; some of these could and were killed by US drones as part of joint military operations.

Cameron's four successors as prime minister, have not had to engage in his innovative legal strategising because, as far as is known, there has been no recurrence of the 'imminent' threat that was perceived to require the killing of Khan. The capacity and willingness of UK and US forces to strike may have had a deterrent effect on terrorist activity, but the more likely impact on terrorist planning was the reduction of the area under ISIL control because of conventional military action in the region. The parliamentary authority to use force in Syria would also have meant that had the UK identified ISIL plotters, then

---

[55] Cameron, D., House of Commons Debate Col. 325 (2 December 2015), https://publications .parliament.uk/pa/cm201516/cmhansrd/cm151202/debtext/151202-0001.htm.

strikes could have been carried out against them without the need to justify them on the basis of an imminent threat. At the same time, there has been a palpable shift in the focus of Cameron's successors away from the threat posed by violent non-state actors towards concerns to do with the implementation and the consequences of Brexit for the United Kingdom, and more recently the return of conventional war in Europe following the Russian invasion of Ukraine. Nevertheless, it is also the case that the RAF has continued to operate against ISIL in Syria on the legal basis of the parliamentary authority that Cameron secured. That is to say the unable and unwilling doctrine, combined with the authority in Resolution 2249.

The delivery by the Attorney General Jeremy Wright, of his speech to the IISS in January 2017 represents continuity of innovating legal thinking between Cameron and May. The timing of the speech is also indicative, however, of the UK government's ambition to form an even closer relationship with the United States post-Brexit. Coming six and a half months after the referendum and nine days before Trump's inauguration as President, this speech was clearly an attempt to signal solidarity with the US government, particularly the defence and intelligence bureaucracies, ahead of the transition to a new US administration. While there is no evidence to suggest that the UK government has retreated from the legal position set out by Wright in his 2017 IISS address, there have been no further formal articulations of this kind. Although the UK government did not applaud the Trump administration's decision to kill General Soleimani (see the Introduction), there is no evidence that this reflected legal concerns about how imminence was being used. Indeed, the Foreign Secretary Dominic Raab stated that 'the right of self defence clearly applies'. Despite this, political and strategic concerns about the damage the attack might do to wider stability in the region, and especially concerns about its impact on relations with Iran, led Raab to call for 'a de-escalation of tensions in the Middle East to avoid war in the region'.[56]

The motif of successive prime ministers has been to focus on the UK's commitment to a rules-based order, including a legal one, as seen in the Johnson government's 2021 'Integrated Review of Security, Defence, Development and Foreign Policy'.[57] Given the highly contested nature of the legal innovations advanced by the UK government to justify its use of force in a counterterrorist role, whether by drone or other military means, domestic critics of these new uses

---

[56] MacLellan, K. Raab calls for de-escalation of tensions after killing of 'menace' Soleimani. *Reuters*, 5 January, www.reuters.com/article/uk-iraq-security-raab-idUKKBN1Z407N.

[57] Global Britain in a Competitive Age The Integrated Review of Security, Defence, Development and Foreign Policy, March 2021, https://assets.publishing.service.gov.uk/government/uploads/system/uploads/attachment_data/file/975077/Global_Britain_in_a_Competitive_Age-_the_Integrated_Review_of_Security__Defence__Development_and_Foreign_Policy.pdf.

of force, whether in parliament, academia, NGOs, and the media have been quick to highlight the disjuncture between the UK government claiming to be a champion of international law, while at the same time employing new legal justifications that they see as weakening the rules-based order. The UK government's use of innovating legal strategies in this regard contrasts sharply with the French government's legal strategising which is the subject of the next section.

2 France: A Strategy of Legal Conventionalism to Meet a Changing Threat Environment

The French government has a long tradition of using military force on the African continent. Since 9/11, it has engaged in counterterrorist uses of force in the Sahel. But there is no equivalent so far of the UK's legal justifications for the killing of Khan nor is there any *opinio juris* supporting a broadened conception of imminence. What has been demonstrated instead by publicly stated French legal rationales is a strong attachment to a strategy of legal conventionalism as part of a distinctly 'French model' (Vilmer, 2021: 10; see also Lushenko et al., 2022b: 5–6) of the use of armed drones. French legal strategising has been shaped by four factors: (i) the changing nature of the threat environment and the associated risk to French citizens and interests; (ii) the articulation of an independent position to that of the United States; (iii) a French approach that seeks UN Security Council authorisation for its uses of force where possible in order to maintain constraints on the use of force more broadly; and (iv) contributing to the maintenance of international security through its military capabilities, especially in the context of upholding its responsibilities as a permanent member of the UN Security Council.

The section shows how the French government has carved out a middle path in its use of force between its desire to stay within a publicly articulated framework of legal conventionalism while progressively embracing the technological possibilities of the drone to respond to the changing threat environment at home and abroad. In navigating this middle path, the French government as we show in the first two parts of the section has been involved in uses of force in both Africa (Mali and the wider Sahel) and the Middle East (Iraq and Syria). The final part of the section is more speculative and explores how the French government might legally justify any future practices of non-battlefield targeted killing, a context that might develop in Mali where the government has withdrawn consent for France and other European states to operate their drones and other aircraft in Malian airspace.[58]

---

[58] Lionel, E. Mali Junta clips German military drone wings, *J News*, 22 November 2022, www .military.africa/2022/11/mali-juntan-clips-german-military-drone-wings/.

The French Model of Drone Operations

Unlike the UK and US modus operandi of the MQ-9 Reaper drones, the 'French model' purposely locates the drone crew (pilot, sensor operator, tactical operator, and image interpreter) in the theatre of operations.[59] This, it is argued, provides better tactical awareness of the local operating environment, allowing better integration and coordination with Special Forces and operational aircrew. In basing its drones in theatre in this way, the French government rebuts the charge, often levelled against US remote warfare, that because such operations are physically separated from the experience of combat, they are risk-free. In adopting this approach, the French government has been keen to stress that its use of drones is in conformity with traditional French military deployments in pursuit of its national interests. For the French, there is nothing novel about drone use, and their operational model is designed to demonstrate this (Vilmer, 2021).

Mali and the Sahel

The French government has employed combat forces, combat aircraft, Special Forces, and drones in countering insurgent forces in the Sahel. Unlike the UK government that has been operating the MQ-9 Reaper in a combat role since 2007, the French were slower and more reluctant in both procuring and deploying Reapers in a strike role (Vilmer, 2021; see also Vilmer, 2017). Following their acquisition of Reapers from the United States in 2013 these were used in an ISR role from 2014, guiding Mirage 2000 aircraft armed with laser guided bombs, and attack helicopters, to their targets. This disaggregation of the ISR/strike process was a political decision borne out of a desire to maintain an independent position and not to be seen as using drones in the same way as the UK and US governments. The French were also reluctant to arm their drones, initially out of concern that civilian casualties caused by French drone strikes could lead to the loss of domestic public legitimacy for their wider operations in Africa (Lushenko et al., 2022b). In addition, there were also concerns over the degree of ongoing control that Washington demanded over MQ-9s sold to its allies. According to Jean-Baptiste Vilmer, former Director of the Institute for Strategic Research (IRSEM) at the French Ministry for the Armed Forces, Reaper drones could not be moved without US permission, and the signals intelligence and payload associated with drone operations had to be American (2017).

---

[59] The French approach contrasts with the British and American *modus operandi* where the only drone personnel in the theatre are the ground crew who maintain and arm the aircraft and those pilots who are present to ensure safe take offs and landings.

However, in response to the changing threat environment, French public opinion favoured arming the Reapers, with a poll conducted in April 2017 showing 74 per cent of the public supportive of the move.[60] As a result of this increased public pressure, the Senate issued a report in May 2017 recommending that the Reapers be armed.[61] Defence Minister Florence Parlay announced on 5 September 2017 that the decision had been taken to arm France's twelve Reapers (Vilmer, 2017).[62] Although that decision was taken in 2017, the armed configuration was not operationally available until December 2019. Within two days of completing final tests, the French Reapers were first used against insurgent forces in Mali to respond to an ambush by the Macina Liberation Front. The French were able to operationalise a strike role for the Reapers quickly because they had been using them since January 2014 in an ISR role in the Sahel. The speed with which the new capability was integrated into operations with existing platforms, however, demonstrates how valuable this new strike power was to the counterterrorist operations being conducted.

For the French government, the addition of armed drones to their inventory was an embrace of an enabling military capability, but so far this has not led to publicly declared innovating legal strategy as in the British case. French policymakers and their legal advisors have relied on two key conventionalist legal justifications for their contribution to counterterrorism operations in Mali. Paris's legal justifications have been the consent of the Malian government, and the legal mandate provided by United Nations Security Council Resolution (UNSCR) 2085. This was adopted unanimously in December 2012 and gave the African-led International Support Mission to Mali (AFISMA) a mandate to assist the government in Bamako 'in recovering the areas in the north of its territory under the control of terrorist, extremist and armed groups and in reducing the threat posed by terrorist organizations'.[63] In the context of

---

[60] Ministère de la Défense, *Baromètre externe « Les Français et la Défense,* Defense Information and Communication Delegation, May 2017, http://data.over-blog-kiwi.com/1/11/98/19/20170513/ob_cfcb92_barometre-externe-de-la-defense-et-d.pdf.

[61] Groupe de travail Les drones dans les forces armées – Examen du rapport d'information (2017). *Commission des affaires étrangères, de la défense et des forces armées : compte rendu de la semaine du 22 mai 2017.* www-senat-fr.translate.goog/compte-rendu-commissions/20170522/etr.html?_x_tr_sl=fr&_x_tr_tl=en&_x_tr_hl=en&_x_tr_pto=sc&_x_tr_sch=http#toc2.

[62] 'France to start using armed drones', AFP.com, 5 September 2017, www.thelocal.fr/20170905/france-to-start-using-armed-drones/.

[63] UN Security Council Resolution 2085, UN Doc S/Res/2085 (2012), 20 December 2012, https://unscr.com/en/resolutions/doc/2085. This was followed on 25 April 2013 by a further Security Council Resolution (2100) that established The United Nations Multidimensional Integrated Stabilization Mission in Mali. Although this Resolution emphasised the 'sovereignty, unity and territorial integrity of Mali', it was adopted under Chapter VII of the Charter and 'Authorizes MINUSMA to use all necessary means [the post-Cold-War euphemism for the use of force],

Resolution 2085, the then Malian President, M. Dioncounda Traore, wrote to President Hollande asking for French military assistance to combat insurgent forces in Mali (Lynch, 2013). On the same day, France's Permanent Representative to the UN, Gérard Araud, submitted a letter to the UN Secretary-General and the President of the Security Council explaining that French military assistance to the Malian government was 'in conformity with international law' and that it would 'last as long as necessary'.[64]

Operation SERVAL was limited spatially to Malian territory, but its follow-on Operation BARKHANE that began in July 2014 included all the 'G5 Sahel' countries – Burkina Faso, Chad, Mali, Mauritania, and Niger (Charbonneau, 2017) and formally ended in November 2022. As part of Operation BARKHANE, France deployed 3,500 troops, 17 helicopters, 4 Mirage attack aircraft based in Chad, 5 MQ-9 Reaper drones, and 2 French-made Harfang drones, these systems being operated only in an ISR role (Dworkin, 2015). President Hollande justified the expansion of military operations, including the use of drones, across the region on the grounds that 'When the Sahel is threatened, Europe and France are threatened' (quoted in Waddington, 2014). The need to respond to the terrorist threat inside Mali that had transnational links across the Sahel led the French government into an open-ended military commitment. However, Paris also believed that it needed to expand the scale of Operation SERVAL to encompass the wider region. French military forces, even after the formal end of Operation BARKHANE, remained in the Sahel region operating out of the French base in Niger.

What has been seen in French military operations in Mali and the wider Sahel has been an expansion of the use of force for counterterrorist purposes that paralleled the UK government's non-battlefield targeted killing discussed in the previous section. However, the divergence between Paris and London is that the former has hitherto relied on a strategy of legal conventionalism – the authority of the UN and the consent of the territorial state – to justify its uses of force whereas the latter has utilised innovating legal strategies – non-temporal conception of imminence and the unable and unwilling doctrine (compare Margot-Mahdavi, 2023: 101–2). This divergence is further manifested in the legal justifications employed by the French and British governments in their projection of force over Syria.

---

within the limits of its capacities and areas of deployment, to carry out its mandate … and requests MINUSMA's civilian and military components to coordinate their work with the aim of supporting the tasks outlined in paragraph 16 above' documents-dds-ny.un.org/doc/UNDOC/GEN/N13/314/17/PDF/N1331417.pdf?OpenElement.

[64] Letter to the President of the Security Council from the Permanent Representative of France to the UN, UN Doc S/2013/17, 11 January 2013, https://undocs.org/pdf?symbol=en/S/2013/17.

Syria

Since 2014 France has been militarily engaged against ISIL in Iraq and this expanded in 2015 to include attacks against ISIL in Syria itself. The French government initially began airstrikes against ISIL targets in Iraq in response to the Iraqi government's request for international assistance. In June 2014, as ISIL seized control of Mosul, the Iraqi government submitted a formal letter to the UN Secretary-General calling 'on Member States to assist us by providing military training, advanced technology and the weapons required to respond to the situation, with a view to denying terrorists staging areas and safe havens'.[65] In response to this request, the French government joined with the UK and US governments in conducting airstrikes against ISIL forces in Iraq. Here, as with its operations in the Sahel, the French government justified its use of force in legal conventionalist terms as an act of collective self-defence in support of Iraq (Recchia and Tardy, 2020; Lushenko et al., 2022b).

The French government extended its military operations against ISIL to its Syrian bases in September 2015 in response to the evolving security situation in the region. Paris carried out its first strike in Syria on 27 September 2015, striking an ISIL training camp (Chrisafis, 2015). Hollande justified the action as a response to ISIL 'extending its grip across the Middle East for the past two years ... from there attacks are organised against many countries, including our own'.[66] The French government reported its use of force to the UN Security Council in a letter dated 8 September 2015. Its letter declared that, 'in accordance with Article 51 of the Charter of the United Nations, France has taken actions involving the participation of military aircraft in response to attacks carried out by ISIL from the territory of the Syrian Arab Republic'.[67]

The French government escalated its military actions in Syria after the 13 November 2015 attacks by ISIL in Paris that killed 130 people. In response to these attacks, Hollande stated that 'what happened yesterday in Paris ... was an act of war. Faced with war this country has to take appropriate decisions. An act of war was committed by a terrorist

---

[65] Letter to the President of the Security Council from Iraq's Permanent Representative to the UN, UN Doc S/2014/440, 25 June 2014, https://undocs.org/S/2014/440.

[66] Bamat, J. 2015. France considers airstrikes against IS group in Syria, rules out ground troops, *France 24,* 7 September,www.france24.com/en/20150907-france-air-strikes-reconnaissance-flights-syria-hollande-press-conference-military.

[67] Letter from the Permanent Representative of France to the United Nations, 8 September 2015, S/2015/745, www.securitycouncilreport.org/atf/cf/%7B65BFCF9B-6D27-4E9C-8CD3-CF6E4FF96FF9%7D/s_2015_745.pdf.

army . . .. France will have no mercy against the barbarians of Daesh'.[68] The French government responded to this assault on its capital by launching, in conjunction with the United States, a series of retaliatory strikes on ISIL targets in Syria that were justified in the words of foreign minister, Laurent Fabius, as acts of 'self-defence'.[69]

One notable feature of the French government's legal position with regard to the conflict in Syria is the complete absence of any reference to the innovating 'unable or unwilling' legal justification that had been pressed into service by the UK and US governments to justify their uses of force in Syria. For Paris, the resort to individual and collective self-defence was sufficient legal cover for their actions, without the need to mobilise the politically more controversial 'unable and unwilling' rationale. However, mindful of alliance considerations, Paris has not publicly criticised the UK and US governments for their mobilisation of this innovating legal justification. Nor has this stopped the French from collaborating in the field with US forces operating under a different legal framework. We would argue that French acquiescence to the 'unable and unwilling' legal claim proffered by the US and UK governments should not be interpreted as *opinio juris* supporting a new customary rule of international law.[70]

Exceptionalism, Silence, or Innovation? French Legal Strategising beyond Sovereign Consent

Given France's commitment to legal conventionalism in its operations in the Sahel and Syria, the question arises how French decision-makers would seek to justify their use of force against a perceived terrorist threat in a context where there was no host state consent or UN Security Council authorisation (non-battlefield targeted killing). This question has been given added salience by the evolving situation in Mali, with the withdrawal of host nation support for French and wider coalition air operations in May 2022 (Margot-Mahdavi, 2023: 101). If French airpower continues to be used against terrorist groups inside Mali, how will Paris legally justify this? One intriguing possibility suggested by Margot-Mahdavi is that French policymakers and legal strategists would refrain from providing any kind of legal justification for such acts, preferring what she has described as

---

[68] 'President Hollande calls Paris attacks an "act of war" – video', *The Guardian*, 14 November 2015, www.theguardian.com/world/video/2015/nov/14/president-hollande-paris-attacks-act-of-war-video.

[69] Quoted in Shaheen, K., Raqqa activists reveal details of French airstrikes on Syria, *The Guardian*, 16 November 2015, www.theguardian.com/world/2015/nov/16/raqqa-activists-french-airstrikes-syria-isis-paris-attacks.

[70] Compare Byers (2003: 181) who argued that customary international law develops either through 'widespread support or acquiescence'.

a position of 'silence' (Margot-Mahdavi, 2020, see also 2023: 101–2). As we have argued, it is rare for states to eschew the need for legal justification in international society, but Margot-Mahdavi is suggesting that the domestic need of the French Executive to avoid any kind of debate, parliamentary or otherwise, over acts of non-battlefield targeted killing is a stronger impulse than the need to reconcile these actions with international legality.

The French government could seek to justify any future non-battlefield targeted killings inside Mali on the grounds that the transboundary threat of terrorism emanating from Mali creates a right of collective self-defence of the other G-5 states. Such an argument would have greater weight had it continued to be backed up by Security Council Resolution 2640 that was adopted under Chapter VII on 29 June 2022 and which authorised MINUSMA to use 'all necessary means' to fulfil its mandate. However, the Transitional Government of Mali's decision to end MINUSMA's mission led the Security Council to end MINUSMA's mandate under Resolution 2640 in June 2023.[71]

The loss of Resolution 2640 as legitimating cover for any future French strikes inside Mali in the context of the withdrawal of Malian state consent poses a potential challenge for French policy-makers who wish to remain within the confines of legal conventionalism. What type of legal strategising might be employed by a French government that found itself confronted with similar circumstances to that which faced the Cameron government in August 2015 in relation to the Khan strike? As with the UK government, the belief that there is a compelling threat to the security of French citizens that needed urgent preventive military action would push the French into new legal territory. Given Lushenko's contention concerning 'French citizens [preference for] juridical strikes' (2022: 5), it is hard to imagine a French government resorting to a strategy of silence in such a context, especially given France's declaratory commitment to a rules-based legal order. What is more, even if French policy-makers are willing to shield a practice of non-battlefield targeted killing that is confined to Mali and the Sahel from domestic and international publics, it is hard to see how they could sustain a strategy of 'silence' in a context where France was using air strikes to attack terrorists that were believed to be planning an attack against the French homeland.

In a situation where the French government lacked host state consent and UN authorisation, it would face two uncomfortable legal possibilities if another allied state did not step in and execute the mission. The first, and we think least likely, is that French policymakers and their legal advisors would follow the UK

---

[71] 'Security Council ends MINUSMA mandate, adopts withdrawal resolution', UN Peacekeeping, 30 June 2023, https://peacekeeping.un.org/en/security-council-ends-minusma-mandate-adopts-withdrawal-resolution.

precedent and reach for a non-temporal conception of imminence to justify the use of force. Two reasons can be adduced here: first, the domestic pressure to ensure that the use of force is supported by proper legal authority and second, the concern about setting still further precedents for an expansion of the unilateral right of self-defence that others might follow.

The second possibility is that the French government might adopt a strategy of moral exceptionalism where no legal justifications are offered and the defence of the action rests solely on moral grounds. This idea has been floated by Vilmer and it was his concern about the risks of precedent-setting that led him to explore this route, drawing on the case of NATO's legal justifications for intervention in Kosovo. In this scenario, French policymakers would not seek to legitimate their use of force in legal terms; instead, they would claim that this was an exceptionalist response that was essential to preventing an impending terrorist attack against French citizens. Significantly, a strategy of exceptionalism of this kind, by contrast with an innovating one, does not seek to establish new legal precedents. In June 2015, Vilmer claimed that, 'In exceptional cases we might be led to conduct targeted killings outside a recognised armed conflict'. He went on to argue that such an action should be justified as an 'exception that recognises the illegal character of the action whilst justifying the exceptional violation of law in certain operations. It must always be stressed that such exceptions do not create any precedent' (Vilmer, 2015). Vilmer articulated this position in contrast to what he perceived as the UK government's strategy of only using the RAF's Reaper drones in a recognised armed conflict which he suggested 'might not be enough' to meet the terrorist threat (2015). As discussed in the previous section, only two months after Vilmer's implied criticism of the limits of the British position, London used the Reaper drone in a non-battlefield targeted killing in Syria.

However, in contrast to Vilmer's 'approach by exception', the UK government claimed that it was acting in conformity with existing law, even if, as we have argued, it was deploying a strategy of legal innovation. What Vilmer is seeking to advance is a position that would reconcile the French government's commitment to upholding a rules-based legal order with a policy that would enable Paris to use force for non-battlefield targeted killing. The key for Vilmer is developing a strategy that condones an 'exceptional violation of law' but does 'not create any [legal] precedent' (Vilmer, 2015).Vilmer cited NATO's 1999 intervention in Kosovo as an example of this kind of moral exceptionalism, and it is true that alliance members, including the United States and France, did not expressly justify their use of force in Kosovo as lawful (the UK government being the singular exception to this). Yet, it is also the case that neither Paris nor any other NATO states claimed that they were breaking the law on grounds of moral exceptionality.

Governments seldom admit to violating the law, recognising both the reputational damage this causes and weakening of the general constraints provided by international law. If actions are taken that are not generally accepted by others as legally compliant, it risks setting new moral and political precedents that others might emulate to justify similar actions. If the French government were to follow Vilmer's 'approach by exception' in justifying future non-battlefield targeted killings, this would risk, contrary to their intentions, pushing out still further the range of legitimating reasons for governments to have recourse to the use of force.

Conclusion

The evolution of both French military action and related legal strategising in the post 9/11 period is instructive of the way Paris has grappled with the competing desires to stay within the confines of a strategy of legal conventionalism and at the same time to rise to the challenges of the new threat environment and technological possibilities afforded by armed drones. Despite sharing the UK's threat perception in relation to ISIL and recognising, like the British, the military opportunities afforded by drones, French policymakers and their legal advisors have been able to meet French security requirements without engaging in the legal innovating of their British counterparts. Nor has the French government provided any express legal support to the innovating legal strategising of its US and UK allies.

The French government's military innovation with regard to armed drones has come in the realm of operations in the way it has deployed its drone crews to the theatre of operations. This shows Paris's desire to avoid the standard political and moral criticisms of 'remote warfare'. For the French their principal area of drone operations has been in Africa, part of a long-standing and continued focus on projecting military power to provide security to and promote stability in the francophone world. The French role in North Africa stems both from its historical relationship with the region and its sense of international responsibility as a great power (Recchia and Tardy, 2020: 7).

In order to ensure effective military intervention in places such as Mali and the wider Sahel, the French government found it prudent to increase the scope of its military assistance by extending both the geographical reach and the temporal range of its activities. The increased tempo and scale of the terrorist attacks within France and elsewhere both cemented public support for such a stance and strengthened the case for action at the governmental level.

The French government's decision to respond positively to Iraq's request for military assistance against ISIL with its use of fighter aircraft against ISIL targets inside Syria was another manifestation of this response. For France, these strikes, like all French uses of force since 9/11, have been guided by a strategy of legal conventionalism. The section has shown how the operational character of the 'French model' of armed drones is paired with a commitment to multilateral authority in the form of UN Security Council authorisation.

The French commitment to legitimating its use of force in this way has three drivers: the first is the need to secure domestic political support because French approval for the use of military force is increased significantly when it is perceived as lawful and the best measure of this is UN Security Council authorisation (Lushenko et al., 2022b). Second, a number of commentators have argued that French governments, particularly after the 1994 Rwandan genocide, has sought multilateral authorisation through the UN to avoid charges of neocolonialism. The concern being that French 'unilateral interventions could be exploited by nationalists within the target state and in neighbouring countries, fuelling anti-French sentiment and undermining France's regional influence' (Recchia and Tardy, 2020: 478). Third, French governments have been reluctant to endorse any use of force that has not been authorised by the UN Security Council, and this stems from two related reasons. The first is that Paris has not wanted to weaken the authority of the Council given that being a permanent member with the veto-bearing power that this entails gives France a privileged position in international society, reinforcing its claim to great power status. Second, French governments since 9/11 have been sensitive to proffering legal justifications for the use of force that weaken UN Charter constraints.

Nevertheless, it is particularly noteworthy that Paris has not publicly objected to the innovative interpretations of international law as articulated and practised by its allies. Given that these challenge the established legal order that France sees itself as a custodian of, it might be expected that it would have mounted a challenge to the UK and US positions as it did over the legal justifications the latter employed in defence of their use of force in the 2003 Iraq war. Mindful of just how politically divisive this was to the transatlantic community, it is perhaps understandable that the French government has been reluctant to reopen these fissures at a time of increased concern about the longevity of the US commitment to European security in the face of a resurgent Russian threat and a return to great power competition more generally. Another important consideration for French policymakers was a desire to limit further friction among allies in the wake of the UK's decision to leave the EU.

French policymakers have found themselves seeking to balance their commitment to a rules-based international order, and Paris's privileged position within it, with the hard-headed security imperative of not alienating the United States. For France, on the one hand, there is the constant desire of not being too dependent on the United States for its security and on the other the perennial search for an independent defence and security identity to avoid being overly dependent on US power in European security. This long-standing ambiguity in French policy has manifested itself in the procurement, joint operations, and targeting decisions surrounding the French Reaper force. At the same time, this dependence on US technological capabilities and operational networks for the use of armed drones has not been matched in the legal strategising employed by French policymakers and their legal advisors. The 'French model', then, is one that is very different to the British one and as the next section shows, it is also distinctive from the approach adopted by Germany.

3 Germany: Legal Conventionalism

Among the major European states, the German government's position on the use of force, and in particular the use of force by armed drones, is the most restrictive. Given the dark shadow cast by its history in the first half of the twentieth century, the German state enshrined in Article 26 of its 1949 Basic Law (the constitution of the Federal Republic of Germany) a clear prohibition on 'aggressive war', while Article 25 of the Basic Law recognises the 'primacy of international law' over Federal Laws. As a result, successive German governments have adopted a strong legal conventionalist approach, only supporting military action where it has a strong basis in international law. At the same time, Germany is a key member of NATO and has relied upon the US security guarantee since the start of the Cold War. Balancing its commitment to upholding international law with its responsibilities as a steadfast ally has posed challenges for German governments, most recently with regard to drone policy and the issue of non-battlefield targeted killing.

The German debate over the parameters for the use of force has been a highly politicised one since the US military response to 9/11. The intensity of this debate has waxed and waned over the period since 9/11, with it being particularly intense between 2010 and 2015 as a result of the large number of US drone strikes and the fact that some of these killed German citizens in the Middle East. With the decline of the prevalence of such actions, and the replacement of terrorism by concerns over migration, then Covid, the issue of non-battlefield targeted killing has become less controversial in German politics.[72] This was reflected in the new Coalition Agreement in November 2021 which endorsed

---

[72] Interview with Andreas Schüller, Programme Director International Crimes and Accountability, European Centre for Constitutional and Human Rights, 19 August 2022.

the need for the acquisition of armed drones for force protection purposes. This was made conditional on both the need for parliamentary approval and Germany maintaining a ban on military participation in non-battlefield targeted killings.

Despite the public political debate over armed drones fluctuating in intensity, a legal debate has been running for several years over the legality of the German government permitting Washington to use the US air base at Ramstein in Germany to facilitate the US policy of armed drone strikes. The section argues that this complex and highly politicised debate in Germany is a key limiting factor on the possibilities for developing any public consensus at a European level on the principles that should guide the use of force. And yet at the same time, the German government feels the need to balance its domestic sensitivities over using force outside of traditional battlefields with its core commitment to European security through its alliance with the United States in NATO.

The section proceeds in three parts. First, we examine the recent protracted domestic political debate over armed drones, mapping the evolution of that debate to the formation of the 2021 Coalition government and the agreement in the Bundestag in April 2022 to arm Germany's drones. Next, we show how the German government combined a strategy of legal conventionalism with limited legal innovation in its decision to operate in a military support role as part of coalition air operations in Syria between 2015 to 2022. Here, we highlight how the political discomfort and legal disquietude of the Green Party (who have held the Foreign Ministry since November 2021) has led to Germany ceasing to participate in this operation from January 2022.[73] The final part of the section highlights the legal tensions that Berlin has had to navigate between alliance commitments on the one hand, in the form of allowing the use of the Ramstein Air Base as a satellite relay station for US drone operations for non-battlefield targeted killing, and, on the other hand, the need to be seen to be in compliance with its own domestic and international law as interpreted by the German courts.

The German Debate

The history of the debate over the procurement of armed drones in Germany is inextricably linked to their controversial use, demonstrating the political sensitivity over all aspects of this technology. So universal has the topic of 'targeted killing' become that all political parties are united in their opposition to using

---

[73] 'Speech by Foreign Minister Annalena Baerbock in the German Bundestag on the extension of the mandate to counter IS', Reuters, 14 January 2022, www.auswaertiges-amt.de/en/newsroom/news/anti-is-mandate/2506642.

force beyond the confines of established international law. Because drones first came to public attention in this context every subsequent discussion of this technology was seen through this particular prism.[74]

Due to this sensitivity then German Defence Minister, Ursula Von der Leyen (now President of the European Commission), told the Bundestag in a debate on 2 July 2014 that:

> The federal government categorically rejects extrajudicial killings which are contrary to international law. This applies to any weapon system …. Our rejection stems from known cases, in which drones that are piloted from a large distance are used for targeted killings of individuals, accepting that innocent persons are hurt. This has nothing to do with the requirement of the Bundeswehr that we are discussing now and in the future (quoted in Cvijic et al., 2019, p. 9).

Nevertheless, von der Leyen's condemnation of 'extrajudicial killings' did not prevent the coalition government she was a member of (2013–2018) from agreeing in 2013 that it would acquire armed drones for force protection purposes.[75] Such a move was met by opposition from the left-wing Die Linke Party as well as from the Green Party who submitted a formal motion against the decision. As a result, procurement was delayed, despite von der Leyen suggesting that the use of armed drones could be subject to a special, higher level of parliamentary approval in order to smooth the path for their acquisition (Franke, 2017: 225).

In February 2018, the new Coalition government decided to lease five unarmed Israeli Heron TP drones for ISR purposes. They did this on the understanding that the decision over whether or not to arm the drones would be left to parliament at a future date. In December 2020 a planned vote in the Bundestag led the SPD to signal its support for armament. However, immediately prior to the vote, the SPD leadership reversed course, withdrawing its support thereby preventing Chancellor Merkel's coalition from arming the Herons.[76]

Successive German governments have been typically risk-averse and reluctant to put their troops in harm's way, and so the decision to refrain from acquiring a military capability that could provide significantly enhanced levels of protection for German forces was perplexing. The CDU was quick to

---

[74] Further, due to a linguistic coincidence, as Ulkrike Franke points out, 'the German term "Drohne" (drone) and "drohen" (to threaten, menace) are close' (See 2017: 197).

[75] Hudon, A. and Boyle, J. German armed forces to acquire armed drones: Defence Minister, *Reuters*, 1 February 2013, www.reuters.com/article/us-germany-drones/german-armed-forces-to-acquire-armed-drones-defence-minister-idUSBRE9100JR20130201.

[76] Staff writer, 'German government at odds over armed drones', *France24*, www.france24.com/en/live-news/20210101-german-government-at-odds-over-armed-drones.

highlight this contradiction in the drone debate and oppose the SPD's position. Then Defence Minister, Annegret Kramp-Karrenbauer, claimed that if the German army does not acquire armed drones, 'we are negligently putting the lives of our soldiers at risk'.[77] Indeed the SPD's reversal was also criticised from within, with then SPD Defence spokesperson Fritz Felgentreu resigning in protest.[78]

The reason why the issue of arming German drones is so controversial in the intra-parliamentary debate is the inextricable association between drone operations and US 'targeted killings'. The SPD leadership also saw political advantage in avoiding an association between themselves and arming German forces with the tools of targeted killing. Beyond these political manoeuvrings, there is also the argument that if the German military ever began to operate armed drones, even if this was initially only for force protection, it would lead to a slippery slope in which drones could be used for counterterrorist purposes in contested legal circumstances. A key part of the concern here is that current and future German governments would be enabled by the technological possibilities of the drone (as we argued in Section 1 was the case with the UK government) to contemplate new actions and missions, thereby lowering the inhibitions to the use of force. In a sense, the German reluctance to embrace drones was out of a concern to avoid the consequences of the technological push that drone technology provides in its capacity to engage different targets in novel circumstances.

In the past few years, the political winds have shifted within the Bundestag on the question of acquiring armed drones. This was no more apparent than in the Coalition Agreement reached in November 2021 which endorsed the need for Germany to acquire armed drones for force protection purposes. This recognised a significant change of position on the part of the SPD and the Green Party in order to be able to form a new government with the Free Democratic Party (FDP). The same document, as in earlier German statements, caveated this support for the German acquisition of armed drones by reaffirming its position that any future German use of armed drones must always be operated according to international law. As we have argued throughout this Element, such a ringing statement of principle neglects to consider the on-going legal contestation between governments, human rights NGOs, and jurists as to what is permitted in terms of the use of force under the rule of self-defence. It does, however, serve the purpose of being seen to be supportive of a legal position for domestic political purposes.

---

[77] Dempsey, J. 'Why Germany's Soldiers Are Denied Armed Drones', *Carnegie Europe*, 5 January 2021, https://carnegieeurope.eu/strategiceurope/83558.

[78] German government at odds over armed drones', *France24*, www.france24.com/en/live-news/20210101-german-government-at-odds-over-armed-drones.

The consensus across the German political spectrum, the German military, and wider expert community is that the government should treat the uncrewed platforms no differently to conventional aircraft. According to Ulrike Franke, the prevailing assumption within the government and the armed forces is that acquisition of armed drones would not lead to any significant change 'either in terms of the types of operations Germany will participate in (never alone, always in cooperation with NATO allies, and – if at all possible – only with a U.N. mandate) or in how these operations will be fought' (Franke, 2016: 2). In practice, however, the notion that drones are no different to other combat aircraft is dispelled by the sensitivities shown towards drones in the German debate. Even after the German parliament took the decision to arm the German Herons in principle for force protection purposes, no subsequent operationalisation of the capability was adopted during the contemporaneous German participation in the UN 2013 Security Council authorised 'Multidimensional Integrated Stabilization Mission in Mali' (MINUSMA).[79] During this mission the German Herons were only operated in an ISR role.

What makes the German drone debate so complex is that there is no direct read across from it to Germany's wider attitude to the use of force more broadly, especially in the context of coalition air operations. This was evident both in the now terminated NATO operation in Afghanistan, where successive German governments participated in a combat role from 2002 to 2021,[80] and the Bundeswehr's non-combat participation in the US–UK–French–Australian military operations against ISIL in Syria between 2015 and 2022.[81] We focus on Syria below because it best illustrates the constraining power of the law in German legal strategising.

Syria

Like its UK and French counterparts, the German government's position on the use of force has evolved to meet the changed security situation on the ground in Iraq and Syria and in response to the terrorist attacks in Europe, especially in Paris. From 2014, the German government had been involved in aerial surveillance over Iraq (operating out of Jordanian and Turkish bases) as well as training Iraqi security forces. A few weeks after the Paris atrocities, the

---

[79] 'Germany leasing additional Heron drone for Mali operations', *Defense Brief* editorial, 7 May 2021, https://defbrief.com/2021/05/07/germany-leasing-additional-heron-drone-for-mali-operations/.

[80] 'German military completes withdrawal from Afghanistan', Reuters, 29 June 2021, www.reuters.com/world/europe/german-military-completes-withdrawal-afghanistan-2021-06-29/.

[81] 'Syria Conflict: German MPs vote for anti-IS military mission', BBC News, 4 December 2015, www.bbc.co.uk/news/world-europe-35002733.

Bundestag voted on 4 December 2015 to authorise military intervention against ISIL in Syria. The German parliament voted overwhelmingly to send a limited military contingent of Tornado reconnaissance aircraft, a frigate, and midair refuelling capacity to the region. Importantly, it stipulated that German forces would not be involved in direct combat operations such as airstrikes. While this was a significant symbolic political contribution, the German commitment to a non-combat role shows the difficulty that German governments have had with the use of force in general.

The German government advanced four separate legal justifications for their deployment of military force. The first was the claim that it was acting in 'collective self defence',[82] the implication being that the right was being exercised on behalf of Iraq (in line with the UK and French arguments justifying their use of force in Syria).

The second legal justification also appealed to self-defence in the context of Germany's legal obligations to support its EU allies as part of the 2007 EU Lisbon Treaty. Foreign Minister Frank-Walter Steinmeier expressly defended Germany's participation in terms of its obligations under Article 42.7 of the Lisbon Treaty.[83] While President Hollande did not invoke Article V of the NATO Treaty as had been expected after the Paris attacks, he did appeal for European solidarity by invoking for the first time the mutual defence provisions in the Lisbon Treaty. The latter creates a legal obligation on all EU member states to provide 'aid and assistance by all the means in their power, in accordance with Article 51 of the United Nations Charter' to member states subject to 'armed aggression'.[84]

The third legal justification employed by the German government (akin to the arguments also employed by the British and the French) located its Article 51 justification in the context of taking actions pursuant to UNSCR 2249. This legal rationale was explicitly set out by the Deputy German Permanent Representative to the UN in his 10 December letter to the UN Security Council.[85] As noted in Sections 1 and 2, UNSCR 2249 was not adopted under

---

[82] Letter from the Chargé d'affaires a.i. of the Permanent Mission of Germany to the United Nations addressed to the President of the Security Council, 10 December 2015, S/2015/946', www.securitycouncilreport.org/atf/cf/%7B65BFCF9B-6D27-4E9C-8CD3-CF6E4FF96FF9% 7D/s_2015_946.pdf.

[83] Federal Foreign Office (2015) 'The fight against ISIS: German Bundestag approves Bundeswehr mandate', https://Bundestag approves mandate for Syria mission (bundesregierung.de).

[84] 'Consolidated Version of the Treaty on European Union', https://eur-lex.europa.eu/resource .html?uri=cellar:2bf140bf-a3f8-4ab2-b506-fd71826e6da6.0023.02/DOC_1&format=PDF.

[85] Paragraph 2 of the letter stated that, 'The Security Council has confirmed in [Resolution 2249] ... that ISIL "constitutes a global and unprecedented threat to international peace and security" and has called upon Member States to eradicate the safe haven that ISIL has established in significant parts of Iraq and the Syrian Arab Republic. ISIL has carried out, and continues to

Chapter VII and as such did not explicitly authorise the use of force. There were widely differing interpretations among Council members as to whether the US, UK, French, and German governments were on secure legal grounds in their interventions in Syria in 2015 (Akande and Milanovic, 2015). The final legal justification pressed into service by German policymakers in 2015 was the innovating, but controversial one, that the right of self-defence under the Charter included using force against a non-state actor (ISIL) on the territory of a sovereign state (Syria) where that state 'does not at this time exercise effective control'. This argument echoes the 'unable and unwilling' language employed by the UK and US governments to justify their uses of force against ISIL in Syria. However, it is noteworthy that the German government in 2015 employed different language to that of their US and UK counterparts – explicitly not 'unable and unwilling', suggesting that German legal strategising was seeking to avoid giving *opino juris* to UK and US claims as to the legality of the unable and unwilling doctrine.

Nevertheless, the new German position was permissive in that it allowed for the use of force in self-defence against a non-state actor in the territory of a state without that state's consent. The willingness of German policymakers and their legal advisors in 2015 to advance what for them was a novel legal justification pushed out the legal boundaries of intervention beyond where previous German governments had gone. This innovating legal strategy was motivated, in part, by the perceived need to respond to the dynamic nature of the threat and the fear that it could be a future target of ISIL attacks. The other key driver of Germany's expansive interpretation of what was permitted under UNSCR 2249 (compared to what had previously been a more conservative stance such as its restrictive interpretation of Resolution 1441 in relation to the use of force against Iraq in 2003) was the need to show solidarity with both its French, EU, and US allies.

What these legal justifications demonstrate about the German government's approach to the use of force in Syria between 2015 and 2021 is an attempt to maintain a legal conventionalist strategy in the context of a prevailing threat environment that demanded a new German military engagement. Despite the commitment to legal conventionalism, the situation required embracing an element of legal innovation as Germany responded to the call from its closest allies. However, it is important to note that the military response was heavily caveated, preferring a support role rather than engaging in combat. Offering political support for the use of force, without actually doing so itself.

---

carry out, armed attacks against Iraq, France, and other States. These States have acted, and continue to act, by taking measures of self-defence' (www.securitycouncilreport.org/atf/cf/%7B65BFCF9B-6D27-4E9C-8CD3-CF6E4FF96FF9%7D/s_2015_946.pdf.

There was no German domestic consensus on the Bundeswehr's military participation in military operations against ISIL in Syria, and the Green Party opposed the decision at the time on grounds of its legality. Speaking six years later as the newly appointed German Foreign Minister, Annalena Baerbock, explained that her party's opposition was not in relation 'to the mission's goals, but our reservations about its legal basis at that time'. She announced in her speech that the German military would continue to operate in the Middle East, but the focus would now be entirely on Iraq and not Syria. She claimed that this mission, unlike its predecessor in Syria, would be 'on a sound footing under international law'.[86] The Coalition government's decision to end a military operation because it lacked a firm 'legal basis' shows how far a constraining view of the law shaped one set of German policymakers decisions while the same legal framework was used to push out the boundaries of intervention by a previous one.

The Green Party's determination to end the German military mission in Syria because of legal inhibitions has not been matched to date by a similar willingness to question the legal basis on which the US government is operating its base at Ramstein in relation to that facility's key role in US global drone strikes.

Ramstein – Backed into a Legal Corner

The Ramstein air base, the largest American military base in Europe, is both a staging post to the Middle East and a relay base for the electronic signals that US expeditionary drone operations support over long distances. Its use has generated controversy in Germany over the domestic and international legality of facilitating such drone operations from German soil. In 2014, the European Centre for Constitutional and Human Rights (ECCHR), along with the human rights NGO Reprieve, took legal action against the German government on behalf of Faisal bin Ali Jaber who had lost two members of his family in a US drone strike in Yemen in the summer of 2012. The case was heard in 2015 in the Cologne Administrative Court, with the claimants requesting that the German government cease allowing the US military to use Ramstein on the grounds that US drone attacks in Yemen were conducted outside the context of an armed conflict and were therefore unlawful. Lawyers for the plaintiffs argued that the German government had a legal responsibility to end its role in these attacks and to take measures to protect the 'right to life'.[87]

---

[86] 'Speech by Foreign Minister Annalena Baerbock in the German Bundestag on the extension of the mandate to counter IS', Reuters, 14 January 2022, www.auswaertiges-amt.de/en/newsroom/news/anti-is-mandate/2506642.

[87] Cologne Administrative Court, 3 K 5625/14, 27 May 2015, www.justiz.nrw.de/nrwe/ovgs/vg_koeln/j2015/3_K_5625_14_Urteil_20150527.html.

However, the German government refused this framing of the killing of bin Ali Jaber's family members. The government's lawyer made the following arguments. First, it was argued that there was no 'right to sue' because in the words of the defending Attorney, 'You have no reliable information that Ramstein Air Base is being used for drone operations'. Here, the German lawyer claimed that the US government 'has always stated that no drones are commanded or controlled from Germany and that Germany is not the starting point of drone attacks'. What is more it was argued that even if Ramstein was a 'satellite relay station . . . used to carry out drone operations, there would be no obligation to protect [Ali Jabber's family members] since the defendant [the German Government] has no legal basis for intervening against this use under the relevant stationing law'. Indeed, the lawyer argued that it was 'not the task of the defendant to act as a world public prosecutor vis-à-vis other sovereign states. Rather, the USA and Yemen are the two acting and therefore solely responsible states'. Despite having argued there was no 'reliable information' that Ramstein was being used for US drone operations, the lawyer proceeded to argue that, *contra* to the plea of the plaintiff, international humanitarian law was applicable to the case of Yemen because 'there is an internal armed conflict there involving the USA and AQAP'.[88]

The legal materials of the case by themselves did not permit a definitive ruling because each party agreed on the relevant law to apply so the judge was left to decide between two plausible competing legal arguments. In response to the German and US denials that Ramstein was being used for drone strikes in Yemen, Judge Hildeund Caspari-Wierzoch accepted the 'plausibility' of the claim that it was being used for such strikes.[89] However, on the question of whether US drone strikes violated international law as argued by Reprieve and ECCHR, Judge Hildeund Caspari-Wierzoch did not make a firm ruling. Instead, she focused her judgement on the question of whether the German government was in compliance with international law. She found no evidence that the German government had violated international humanitarian law and on this basis the judge rejected the plaintiff's arguments. In her judgement the German government had fulfilled its positive obligation to protect the right to life by seeking assurances from Washington that it complies with international law in conducting its strikes.[90]

---

[88] Cologne Administrative Court, 3 K 5625/14, 27 May 2015, https://www.justiz.nrw.de/nrwe/ovgs/vg_koeln/j2015/3_K_5625_14_Urteil_20150527.html.

[89] Connolly, K., 'Court dismisses claim of German complicity in Yemeni drone killings', *The Guardian*, 27 May 2015, www.theguardian.com/world/2015/may/27/court-dismisses-yemeni-claim-german-complicity-drone-killings.

[90] Talmon, S., 'Federal Administrative Court rules that the United States may continue to use its air base in southern Germany for lethal drone strikes in Yemen', *German Practice in International Law,* 12 October 2021, https://gpil.jura.uni-bonn.de/2021/10/federal-administrative-court-rules-that-the-united-states-may-continue-to-use-its-air-base-in-southern-germany-for-lethal-drone-strikes-in-yemen/.

It is important to note that in reaching this conclusion the judge emphasised that even if 'the extraterritorial binding of fundamental rights is affirmed in principle . . . the foreign policy framework and the provisions of international law must be taken into account in the concrete intensity of protection depending on the individual case'.[91] Even if the German state had a 'duty to protect' then, the judge was emphatic that this did 'not give rise to a right to termination of the treaty on the stay of foreign armed forces in the Federal Republic of Germany of October 23, 1954 (Federal Law Gazette 1955 II p. 253), which permits the stationing of foreign troops in Germany, or the NATO Troop status according to Art. XIX NTS'. Here the judge explained that any such termination would 'inevitably adversely affect numerous other vital and legitimate interests of the Federal Republic in foreign and defense policy cooperation'.[92]

Nevertheless, the judge had sufficient sympathy towards the plight of the Plaintiffs that she took the unusual step of allowing for them to appeal her ruling.[93] The case was heard again in March 2019 before the Senate of the Higher Administrative Court in Münster. By contrast with the 2015 Judgement, the Senate agreed with the plaintiffs that the German government had not fulfilled its positive obligations to protect the right to life, finding that the measures taken to ensure the legality of US strikes were insufficient.[94] The Munster Court questioned how far US political and military authorities were in compliance with their international legal obligations to protect civilians. The Munster Court concluded that, 'There are considerable doubts as to whether the general practice of the United States on the use of armed drones in Yemen adequately takes into account the principle of distinction as required by international humanitarian law; in particular, whether specific attacks are limited to those persons who, as members of a party to a conflict, fulfil a continued combat function or, as civilians, directly participate in hostilities'.[95] The ruling stated that 'Reliable information on drone strikes in

---

[91] Cologne Administrative Court, 3 K 5625/14, 27 May 2015, 3 K 5625/14, www.justiz.nrw.de/nrwe/ovgs/vg_koeln/j2015/3_K_5625_14_Urteil_20150527.html.

[92] Cologne Administrative Court, 3 K 5625/14, 27 May 2015, www.justiz.nrw.de/nrwe/ovgs/vg_koeln/j2015/3_K_5625_14_Urteil_20150527.html.

[93] Connolly, K., 'Court dismisses claim of German complicity in Yemeni drone killings', *The Guardian*, 27 May 2015, www.theguardian.com/world/2015/may/27/court-dismisses-yemeni-claim-german-complicity-drone-killings.

[94] North Rhine-Westphalia Higher Administrative Court, judgment from 19/3/2019 – 4 A 1361/15 – Wording of the oral pronouncement of the Judgment https://ecchr.eu/fileadmin/Juristische_Dokumente/OVG_Muenster_oral_declaration_of_judgment_19_March_2019_EN.pdf (based on translation by ECCHR).

[95] Quoted in Talmon, S., 'Federal Administrative Court rules that the United States may continue to use its air base in southern Germany for lethal drone strikes in Yemen', *German Practice in International Law*, 12 October 2021, https://gpil.jura.uni-bonn.de/2021/10/federal-administrative-court-rules-that-the-united-states-may-continue-to-use-its-air-base-in-southern-germany-for-lethal-drone-strikes-in-yemen/.

Yemen including from official US sources indicates instead that this process of distinguishing, required by international law, is insufficiently carried out, and not just in isolated cases'.[96] As a result, 'the Senate is convinced that the Federal Republic of Germany has a positive constitutional obligation of protection related to life and limb, which stands opposite an entitlement of the plaintiffs that has as yet not been sufficiently met'.[97] However, the Munster Court also was emphatic in its ruling that this did not 'mean that Germany must act to prohibit the use of Ramstein Air Base for drone operations'.[98] The legal responsibility of the German government was to 'work towards compliance with international law by taking measures that it deems to be suitable'.[99]

In yet another turn of the legal wheel, the German government appealed the decision and in November 2020 the Federal Administrative Court in Leipzig overturned the Munster Court's decision. Reviewing the case and the legal materials of the plaintiffs and lawyers for the government, the Leipzig court ruled that there were not sufficient 'factual indications' of actual and on-going violations of international humanitarian law on the part of the United States. As the international lawyer Stefan Talmon expressed it, 'the Federal Administrative Court required for the constitutional duty to protect ... an actual violation of international law on the part of the other State .... The duty to protect ... required the German State to intervene only in situations where there was a general pattern of illegal acts under international law and not just isolated individual violations'.[100] The Federal Administrative Court overturned the Judgment of the Higher Administrative Court because the latter had not provided sufficient factual evidence that the US drone operations in Yemen involving Ramstein Air

---

[96] North Rhine-Westphalia Higher Administrative Court, judgement from 19/3/2019 – 4 A 1361/ 15 – Wording of the oral pronouncement of the judgement, www.ecchr.eu/fileadmin/ Juristische_Dokumente/OVG_Muenster_oral_declaration_of_judgment_19_March_2019_EN .pdf (unofficial translation by ECCHR).

[97] North Rhine-Westphalia Higher Administrative Court, judgement from 19/3/2019 – 4 A 1361/ 15 – Wording of the oral pronouncement of the judgement, www.ecchr.eu/fileadmin/ Juristische_Dokumente/OVG_Muenster_oral_declaration_of_judgment_19_March_2019_EN .pdf (unofficial translation by ECCHR).

[98] North Rhine-Westphalia Higher Administrative Court, judgement from 19/3/2019 – 4 A 1361/ 15 – Wording of the oral pronouncement of the judgement, www.ecchr.eu/fileadmin/ Juristische_Dokumente/OVG_Muenster_oral_declaration_of_judgment_19_March_2019_EN .pdf (unofficial translation by ECCHR).

[99] North Rhine-Westphalia Higher Administrative Court, judgement from 19/3/2019 – 4 A 1361/ 15 – Wording of the oral pronouncement of the judgement, www.ecchr.eu/fileadmin/ Juristische_Dokumente/OVG_Muenster_oral_declaration_of_judgment_19_March_2019_EN .pdf (unofficial translation by ECCHR).

[100] Quoted in Talmon, S., 'Federal Administrative Court rules that the United States may continue to use its air base in southern Germany for lethal drone strikes in Yemen', *German Practice in International Law*, 12 October 2021, https://gpil.jura.uni-bonn.de/2021/10/federal-administra tive-court-rules-that-the-united-states-may-continue-to-use-its-air-base-in-southern-germany- for-lethal-drone-strikes-in-yemen/.

Base constituted a systematic pattern of violations in relation to indiscriminate attacks against civilians. Moreover, the Federal Administrative Court considered, *contra* the Munster Court, that the German government had fulfilled its 'duty to protect' foreigners under both German and international law by seeking and receiving assurances from the US government that any US drone operations involving Ramstein would be in conformity with international law.[101]

What this legal dispute demonstrates is both the contested nature of US drone policy within German politics and the way that the same legal materials were mobilised by the different parties to support their respective claims as to the jurisdiction of the German courts over the legal use of the Ramstein Air Base for US drone operations. What is more, the issue is not closed because the plaintiffs made a still further appeal on 21 March 2021 to the Federal Constitutional Court, the highest court responsible for interpreting the Basic Law. At the time of writing, this case is yet to be heard. What is at stake here is the future of US drone operations out of Ramstein because it is plausible that the Federal Constitutional Court could decide that US operations in Yemen and elsewhere that are supported by the satellite relay station do constitute a sustained pattern of violations of international humanitarian law. As a result, the Federal Constitutional Court could decide that the German government needs to do more to remain compliant with international law which is part (Article 25) of the Basic Law itself. This could potentially include even ending the use of Ramstein as a satellite relay station for US drone strikes. Such a momentous legal decision would have wider political ramifications not only for US drone use in Germany, with the German government also hosting a US base (US Africa Command) that has played a key role in supporting US drone strikes in Somalia, but also for the use of other US bases in Europe such as the Naval Air Station in Sicily which has supported US drone strikes in Libya. For those who interpret US drone strikes in Yemen, Somalia, and Libya as violations of international humanitarian law, such a decision is vital to counter such illegal basing practices in Germany and elsewhere.[102]

---

[101] Talmon, S., 'Federal Administrative Court rules that the United States may continue to use its air base in southern Germany for lethal drone strikes in Yemen', *German Practice in International Law,* 12 October 2021, https://gpil.jura.uni-bonn.de/2021/10/federal-administrative-court-rules-that-the-united-states-may-continue-to-use-its-air-base-in-southern-germany-for-lethal-drone-strikes-in-yemen/. See also 'Ramstein at Court: Germany's role in US drone strikes in Yemen', www.ecchr.eu/en/case/important-judgment-germany-obliged-to-scrutinize-us-drone-strikes-via-ramstein/.

[102] Dwelle, M. E., 'U.S. Base Off-Base? Drone Hub Faces Challenge in German Court', *Columbia Journal of Transnational Law*, 17 April 2021, www.jtl.columbia.edu/bulletin-blog/us-base-off-base-recent-challenge-to-us-air-base-in-germany.

Conclusion

The German case highlights the complex interplay between pursuing a primarily conventionalist legal strategy and maintaining a vital positive relationship with allies. This tension manifests itself in the German government remaining silent in the face of legally contested uses of force by armed drones on the part of the US and UK governments that challenge Berlin's legal conventionalist commitments. While there has been a strong consensus across the German political spectrum that conventionalism entails a rejection of non-battlefield targeted killing, there was disagreement between the political parties over whether Germany's participation in a non-combat role against ISIL in Syria could be legally justified. The Green Party formally opposed the decision of the Merkel-led Coalition government in 2015 to participate in Syrian operations on legal grounds and once in government seven years later, they justified their termination of the Bundeswehr's role because of these concerns. What this illustrates in contrast to the British and French cases is that legal strategising has an important intra-party dimension in the German case, showing once again the contested nature of the law in relation to the use of force.

It is important to appreciate that German policymakers have shared the perception of threat from terrorist groups of their UK and French counterparts. However, this shared threat perception has not resulted in a convergence on legal strategising, with German policymakers emphasising the constraining power of the law. At the same time, there has been a strong collective negative reaction across the German debate regarding the technological possibilities of the drone in terms of the new military capabilities it offers for non-battlefield targeted killing. As a result, far from embracing this technology in the form of the MQ9 Reapers as the British and French have done (as well as other NATO allies, the Netherlands, Italy, Spain) German governments have leased the much less capable Israeli Heron 1 drone, and even then they have so far only operated it in an ISR role. The problem for German governments has not been that of the drone technology per se, but its indelible associations in the domestic political debate with non-battlefield targeted killing symbolised by the US and UK uses of the MQ-9 in a counterterrorist role.

What explains the different character of German policy responses (procurement, deployment, and use) and associated legal strategising compared to the UK and French cases is Berlin's strong aversion to deploying its military in conflicts outside the NATO area. Nevertheless, its position as a beneficiary of the transatlantic security guarantee has required the German government to navigate this potential contradiction by avoiding the articulation of a clear legal position in relation to the legality of US and UK operations involving the use of drones for non-battlefield targeted killing.

The German government's reticence to take a public position on these legal questions is not solely linked to the controversy surrounding the US use of the Ramstein Base for its drone operations. It also extends to non-battlefield targeted killing as was demonstrated by the German government's public silence following the UK strike in 2015 that killed ISIL members. The reason why Berlin has been cautious in not criticising US and UK drone strikes, and the use of Ramstein, is because it wants to maintain its positive relationship with its security guarantor and other NATO members. As a result, the German government has found itself acquiescing in US and UK uses of force, and the innovative legal justifications supporting them.

That said, there were elements of legal innovation in the Merkel-led 2015–2021 Coalition government's legal justifications for the German military's non-combat role in the US-led allied operations in Syria. The use of the 'unable and unwilling' principle, albeit by another name, and the claim that Germany had legal obligations to assist under the EU Lisbon Treaty after the French President had invoked the mutual defence obligation commitment represented a novel interpretation of the self-defence rule. It was novel because Berlin had never made these legal arguments before and it was the first time that a state had justified a military deployment based on European alliance solidarity. Nevertheless, the German contribution was limited in material terms, whatever its political benefits in terms of displaying alliance solidarity, remaining strictly a support one, and no German aircraft were engaged in hostile action.

What the Merkel-led government's participation in the Syrian operation shows is that even an actor committed to a strategy of legal conventionalism may find itself having to balance its response to the legal innovations of its allies in the context of both its enduring strategic alliances and the changing nature of the international security environment. This balancing act, however limited the actual military contribution, was domestically so contentious that its contested legal basis could only be legitimated domestically by a particular coalition of partners for the time they were in office. The successor Coalition government, driven by Green Party legal inhibitions about the operation, viewed the legal strategising as an innovation too far and ended the operation.

Current and future German policymakers and their legal advisors can be expected to justify any future uses of force by Germany and their response to the use of force by others in legal conventionalist terms. The German government has been sensitive about criticising US – and by extension UK – use of armed drones, mindful that there have been growing uncertainties in Berlin about the long-term viability of the US guarantee to European security. Although the Biden administration offered reassurances to its NATO allies about the US commitment to Europe, the reliability of that guarantee was called into question in German and European eyes as a result of the unpredictabilities of the Trump administration and the persistence of the 'America First' agenda within the Republican Party. The importance of the US guarantee to European security has been underlined by Putin's illegal invasion of Ukraine in 2022 and the challenge to the wider European security order that this represents.[103] The German government's decision to double its defence budget in the aftermath of the Ukraine invasion is not just a response to threatening developments to the East, it is also predicated on demonstrating its alliance credentials in what remains a highly polarised US debate about the value of Europe to US security. The future German debate, policymaking, and associated legal strategising in relation to the use of force will reflect a constant juggling between maintaining solidarity with European allies, especially France, ensuring the United States remains anchored to Europe's security in order to deter future Russian aggression, and the need to be seen as upholding core domestic and international legal principles. The complexities of this juggling act explain why German governments have found themselves disagreeing over the legal boundaries of projecting military force beyond the European space, but aligned on the need not to jeopardise alliance relations by expressing public opposition to uses of force that they perceive as legally precarious on the part of those allies. This is a precarious balancing act and one that could potentially become unbalanced – and even be destabilised – if the Federal Constitutional Court finds that the German government is party to US military actions that systematically violate international law. What is clear is that such a judgement will not belong solely to the realm of legal argumentation; instead, the complex interplay of law and politics will shape the possibilities of decision.

4 Attempts to Achieve a European Consensus

Having provided three in-depth case studies of the major European military states, this section turns to the debate concerning efforts at achieving a European consensus on the use of armed drones, and more widely, the use of force. The European debate regarding the legality and legitimacy of the use of armed drones is both fractured and muted, with a general lack of consensus among states revealed primarily through inaction and conspicuous silence rather than by explicit statements and publicly announced positions. Nonetheless, efforts at finding and formalising a consensus on the use of armed drones have taken place within some European Institutions, namely, the European Parliament and the European

---

[103] Dunn, David Hastings., 'Ukraine war: this conflict is just further evidence that Putin's Russia is now a rogue power', *The Conversation*, 20 October 2022, https://theconversation.com/ukraine-war-this-conflict-is-just-further-evidence-that-putins-russia-is-now-a-rogue-power-192972.

Commission. However, despite pressures from these European bodies, deliberations over a new consensus on the use of armed drones are conspicuously absent from the highest level of EU decision-making, the European Council. In this section we chart the limited efforts at developing a new consensus, but show how appeals to a common body of international law have masked divergent interpretations between European states as to the permissible limits of the use of force for non-battlefield targeted killing. As a consequence, discord and fragmentation has characterised the debate among European states, preventing the emergence of any form of consensus.

Seeking Consensus

Efforts at addressing the legality of the use of armed drones at the European level began within the European Parliament in April 2012. Four Members of the European Parliament (MEPs) across a range of parties, published a written declaration on the use of armed drones. The declaration, which came in the wake of a significant surge in the frequency of armed drone strikes under the Obama administration, urged the EU and its member states to 'categorically ... prohibit drone operations for the purpose of targeted killings' and to 'publish their criteria for combat drone operations'.[104] The articulation of this European concern for greater constraints on the use of armed drones reflected a broader push for transparency, accountability, and limitation in the use of force. In 2013, the European Parliament's Subcommittee on Human Rights released a report on the transparency and accountability of US drone strikes. This recommended that the European Council 'launch a broad inter-governmental policy dialogue aiming at achieving international consensus ... on the legal standards governing the use of currently operational unmanned weapon systems'.[105] This call to action was largely ignored and no such inter-governmental dialogue has taken place at member states level. Nonetheless, the report and its call for consensus galvanised further efforts within the European Parliament.

In February 2014, the European Parliament passed resolution 2014/2567 (RSP)[106] calling for a common EU position on the use of armed drones.

---

[104] Lösing, S., Alfano, S., Gomes, A., Tavares, R. (2012) 'WRITTEN DECLARATION pursuant to Rule 123 of the Rules of Procedure on the use of drones for targeted killings', *European Parliament*, www.europarl.europa.eu/sides/getDoc.do?pubRef=-//EP//NONSGML% 2BWDECL%2BP7-DCL-2012-0002%2B0%2BDOC%2BPDF%2BV0//EN.

[105] Quoted in Melzer, N. (2013) 'Human Rights Implications of the Usage of Drones and Unmanned Robots in Warfare', *European Parliament Directorate-General for External Policies of the Union Directorate B Policy Department,* https://europarl.europa.eu/RegData/ etudes/etudes/join/2013/410220/EXPO-DROI_ET(2013)410220_EN.pdf.

[106] European Parliament (2014) 'European Parliament resolution of 27 February 2014 on the use of armed drones (2014/2567(RSP)', www.europarl.europa.eu/doceo/document/TA-7-2014-0172_EN.html?redirect.

The resolution recognised that the use of drones had 'increased steeply over the past decade' and that 'unknown numbers of civilians have been killed, seriously injured or traumatised by drone strikes outside of armed conflict zones'. It further claimed that 'drone strikes outside a declared war by a state on the territory of another state without the consent of the latter or of the UN Security Council, constituted a violation of international law and of the territorial integrity and sovereignty of that country'. Within the resolution, the European Parliament called upon the European Council 'to develop an appropriate policy response at both the European and global level which upholds human rights and international humanitarian law'. Specifically, it invited the High Representative for Foreign Affairs and Security Policy, the Member States and the [European] Council to 'oppose and ban the practice of extrajudicial targeted killings' and 'ensure that the Member States, in conformity with their legal obligations, do not perpetrate unlawful targeted killings or facilitate such killings by other states'. In addition, the European Parliament recommended that the Council 'promote greater transparency and accountability on the part of third countries in the use of armed drones with regard to the legal basis for their use and to operational responsibility, to allow for judicial review of drone strikes and to ensure that victims of unlawful drone strikes have effective access to remedies'.[107]

The increased visibility of this issue in the parliament did not lead to any significant breakthrough at the governmental level among member states. Faced with such a disappointing response, a second resolution was passed by the European Parliament in April 2016. This reiterated the parliament's 'call on the [European] Council to adopt an EU common position on the use of armed drones'.[108] In response to a series of follow-up questions from MEPs advocating for these resolutions, Federica Mogherini (then High Representative of the European Union for Foreign Affairs and Security Policy and Vice-President of the EU) responded that, 'the establishment of a legal framework about armed drones does not rest with the Commission but with EU Member States'. Significantly, she added that 'there is no common position on the issue'.[109] The phrasing of this response is revealing in that it implies a lack of willingness

---

[107] European Parliament (2014) 'European Parliament resolution of 27 February 2014 on the use of armed drones (2014/2567(RSP)', www.europarl.europa.eu/doceo/document/TA-7-2014-0172_EN.html?redirect.

[108] European Parliament (2016) '2016/2662(RSP) Resolution on attacks on hospitals and schools as violations of international humanitarian law', https://oeil.secure.europarl.europa.eu/oeil/popups/ficheprocedure.do?lang=en&reference=2016/2662(RSP).

[109] European Parliament (2016) 'Answer given by Vice-President Mogherini on behalf of the Commission, Question reference: E-005455/2016', www.europarl.europa.eu/doceo/document/E-8-2016-005455-ASW_EN.html.

on the part of member states to engage with this agenda. The reason why governments have been so reluctant to debate this question is three-fold: first, there has been no pressing need to do so; second, they recognise it is a politically contested issue; and third, while governments share a common commitment to upholding international law, there is a recognition that they disagree over its interpretation in relation to the use of force. This sentiment was echoed by a senior member of the European External Action Service, the diplomatic arm of the European Union, who explained that while adherence to international law appears to be a 'common minimum denominator . . . the problem is what does international law mean about the terrorism, counter-terrorism from Iraq to targeted strikes, we know that there is no consensus in the international community and it's not specific to the EU'.[110]

The UK's 2015 ISIL Attack and the Further Fracturing of Any Possible European Consensus

The possibilities for a European consensus on the use of armed drones have been taken up by academics and the wider civil society community. A significant study in this regard was Jessica Dorsey and Christoph Paulussen's April 2015 research paper *Towards a European Position on Armed Drones and Targeted Killing: Surveying EU Counterterrorism Perspectives*. They sent a questionnaire to twenty-eight foreign ministries in EU member states, asking the following questions: (i) whether or not the use of armed drones outside of armed conflict would be legal if the target presents an imminent threat; (ii) whether or not a state can be engaged in an armed conflict with a non-state actor without specific qualifications; (iii) whether a geographically boundless or global armed conflict is legally possible; and (iv) whether or not the use of armed drones globally was generally speaking in conformity with international law? Tellingly, they received only five responses, with only 18 per cent of the questions being completed by those who responded. Notably, there was no consensus on the four questions and the Netherlands, and the Czech Republic (the only two non-anonymous respondents), did not wholly agree on any single one of these questions. What the responses to this survey show is that governments would rather maintain a position of silence, or at best ambiguity on what the law permits in relation to the use of force rather than trying to seek a common position which would open up major fissures between the European states.

The Dorsey and Paulussen paper might have had a better reception, especially in British policymaking circles, had it not been published only a few months before the Cameron government employed a broadened conception of

---

[110] Interview with senior EU official, Brussels, April 2017.

imminence to justify its killing of Khan in Syria. This strike pushed any prospect of consensus even further down the road, closing down the possibility that had existed prior to this when consensus might have been more achievable. The opportunity for the UK government to play a leading role in developing a consensus at the European level had been identified by the Birmingham Policy Commission (BPC) in its 2014 report, *The Security Impact of Drones: Challenges and Opportunities for the UK*. The report identified that British state practice in relation to armed drones had, to this point, rejected the use of force for non-battlefield targeted killing. The Commission, in looking ahead to the security landscape in 2035, projected a world where there was a European consensus on the use of armed drones. The report anticipated that 'the use of armed [drones] by the Central Intelligence Agency (CIA) and USAF for targeted killing outside areas of recognised armed conflict will [have ended], thus bringing the United States and its NATO allies into a common position on the lawful use of armed' [drones].[111] A key assumption of this analysis was the projected reduction in the threat from terrorist and insurgent groups as the post-9/11 wars were expected to come to an end. This assumption, together with the hoped-for consensus, proved to be premature at best. What happened instead, with the outbreak of civil war in Syria, was the expansion of an area of ungoverned space from which terrorist acts have been planned and orchestrated.

The BPC argued that the UK government would have an 'instrumental role to play in creating' a new NATO consensus on the use of armed drones. In practice, however, any prospect of bringing the EU's NATO allies 'into a common position' was significantly undermined by the UK government's shift to legalise non-battlefield targeted killing. The change in British practice as set out in Section 1 was accompanied by a major change in legal justification that complemented US justifications for using force outside of established battlefields. Although its key allies in Europe were careful not to publicly condemn the UK's innovating legalist approach to imminence, the UK government's killing of two UK nationals who were members of ISIL in August 2015 torpedoed any prospect of realising the aspiration of European parliamentarians, NGOs, and academics for a new European constraining consensus on this form of the use of force.

The 2016 UK referendum on membership of the EU set in motion the process of Brexit adding further complications to the notion of a security consensus in Europe. For supporters of a common position on the legality of the use of force, the UK's departure was viewed as an opportunity for greater European foreign

---

[111] The Security Impact of Drones: Challenges and Opportunities for the United Kingdom, Report of the 2014 Birmingham Policy Commission, October 2014, www.birmingham.ac.uk/Documents/research/policycommission/remote-warfare/final-report-october-2014.pdf, p. 8.

and security policy integration within the remaining EU. For them, an EU depleted of the United Kingdom was believed to be more likely to reach consensus on such a controversial security question. For others, however, no meaningful consensus could be formed on the permissible boundaries of the use of force which did not include such a key security player as the United Kingdom. The ambivalent way in which the United Kingdom was viewed in this debate can be seen in the 2021 Chatham House report on *Military drones in Europe*. Nilza Amaral and Jesicca Dorsey argued that despite Brexit, the United Kingdom because of its commitment to democratic values and a rules-based international order, should be involved in any dialogue over best practices governing the use of force. However, it was assumed by these authors that because the UK government incurred criticism from some non-governmental quarters for its non-battlefield targeted killing in Syria in 2015, it would now be willing post-Brexit to be 'brought into the fold' by collaborating with the EU in agreeing new guidance on the rules regulating the use of force. Crucially, Amaral and Dorsey interpreted this to mean a new consensus that could represent a retreat away from the UK's shift 'towards a broader conception of imminence' (Amaral and Dorsey, 2021: 31).

As Brexit has unfolded in practice, however, the United Kingdom has not sought to align its foreign and security policies with the EU. Indeed, the opposite has more often been the case, with successive UK governments seeking to demonstrate their claims that life outside the EU provided new freedoms for doing things differently as 'Global Britain'. A further aspect of Brexit has been the degree to which the UK government has sought an even closer position with the US government rather than its European allies in matters of defence and security. As noted in Section 1, Jeremy Wright's post-Brexit keynote IISS speech was one instrument by which London sought to deepen this connection. For Theresa May, Brexit heralded an opportunity 'to renew the special relationship for this new age. We have the opportunity to lead, together, again'.[112] This trend was even more apparent when Boris Johnson succeeded May as prime minister in 2019 and oversaw an Integrated Review of Security, Defence, Development and Foreign Policy for the United Kingdom in 2021. One significant manifestation of this has been the shift to the 'Indo-Pacific' in UK security thinking.[113]

---

[112] May, T., 'Prime Minister's speech to the Republican Party conference 2017', 26 January 2017, www.gov.uk/government/speeches/prime-ministers-speech-to-the-republican-party-confer ence-2017 (see also Seldon, 2020: 181).

[113] 'Global Britain in a competitive age: The Integrated Review of Security, Defence and Foreign Policy', March 2021, https://assets.publishing.service.gov.uk/government/uploads/system/ uploads/attachment_data/file/975077. The UK Government published in March 2023 'Integrated Review Refresh 2023: Responding to a more contested and volatile world' that

The 2021 security partnership between Australia–United Kingdom–United States (AUKUS) is a conscious effort by the United Kingdom to deepen existing security cooperation between these three states in defence technology, cybersecurity, hypersonic warfare, and AI.[114] A key feature of the new trilateral security partnership was the UK and US government's agreement to transfer nuclear submarine capabilities to Australia as part of countering Chinese power in the region. The establishment of AUKUS, much to the disquiet of the French government which saw Australia renege on a commitment to buy French-made submarines, was a clear statement of the UK government's attempt to reinvent itself as a global, rather than as a European power. An associated aspect of this attempt to be a closer ally of the United States has been that the UK government values its freedom of action to act militarily on a wider canvas more than it does a closer relationship with the EU on these matters.

Prospects for Proliferation

A growing feature of the European and international security landscape has been the spread of drone capabilities to an increasing number of states. In Europe, in addition to the UK, French, and German governments that currently operate drones in an armed role, Italy, Belgium, the Netherlands, Greece, Poland, and Switzerland all operate drones that have a capability to be armed should a political decision be taken to do so. Beyond Europe the proliferation is more widespread and continues to grow.

This proliferation of drone acquisition in Europe and beyond raises the question as to what impact the development of an actual armed capability by this grouping of states, and others in the future, will have on the possibilities for reaching a consensus on the legal interpretation of the rules governing the use of force. For most of these states the acquisition of armed drones has been accompanied by statements outlining that these systems are no different from any other military equipment and, just like these other tools, would be used in conformity with international law. It is plausible that these newly armed drone states will not use force for non-battlefield targeted killing because they are concerned not to lend support to a more expansive interpretation of the self-defence rule. Three reasons can be adduced for this: first, that they want to avoid the accusation of law-breaking that has been levelled by critics at the more

---

confirmed this growing focus on the 'Indo-Pacific' in UK thinking, www.gov.uk/government/publications/integrated-review-refresh-2023-responding-to-a-more-contested-and-volatile-world.

[114] Wintour, P., 'What is the Aukus alliance and what are its implications?', *The Guardian*, 16 September 2021, www.theguardian.com/politics/2021/sep/16/what-is-the-aukus-alliance-and-what-are-its-implications.

permissive approach to the meaning of self-defence espoused by London, Washington, and Canberra; second, that they share the critics' concern that a non-temporal conception of imminence risks setting a dangerous precedent for others to emulate; and third, that they wish to be seen to be working in harmony and solidarity with the major EU states interested in building a common approach to security and defence issues.

Set against this constraining outcome, however, is an alternative possibility. In this future environment, an increasing number of European armed drone users might use their new assets for non-battlefield targeted killing and justify this in terms of a broadened conception of imminence, providing new state practice and *opinio juris* to support the legal position advanced by the US and UK governments. Such an approach would push back against the French and German legal conventionalist position and represent an increasing number of states advocating that customary law should recognise a non-temporal conception of imminence.

The UK experience shows that states that have a restrictive interpretation of the self-defence rule and then acquire a new capability that opens up risk-free possibilities for non-battlefield targeted killing may shift the legal calculus in a more permissive direction. Put differently, while a new capability does not determine the meaning and interpretation of legal rules, it does open up a space in which new actions – and new legal strategies with which to justify these – become possible.

Consequently, what the above shows is that any future movement towards an increased European consensus on legal interpretations will not necessarily lead to a more constrained approach to the use of force. We agree with Amaral and Dorsey that 'having more clarity on legal interpretations could allow for a more legitimate understanding of the use of force potential in situations of counter-terrorism operations' (2021: 38). However, what they do not discuss is the possibility that increased convergence over legal interpretations and justifications could be in support of a non-temporal conception of imminence which they and others oppose.

At one end of the scale, the UK government would like to see a general acceptance on the part of other governments – expressed in both state practice and *opinio juris* – of the legality of their permissive uses of force. Failing this endorsement, London, long used to managing intra-alliance politics, will have to be satisfied with the current situation where there is muted acquiescence on the part of others in Europe. For critics of the UK government's drone strikes, however, a maximalist objective would be a retreat by London from the new conception of imminence and a growing body of state practice and *opinio juris* that publicly challenges those states that continue to espouse a more expansive interpretation of the self-defence rule. The third possible outcome, and which we believe is probably the most likely, is where the present situation persists of neither a consensus on either a constraining or permissive interpretation of the self-defence rule.

Conclusion

The use of armed drones to make military strikes against suspected terrorists in both recognised armed conflicts and non-battlefield settings has been a constant source of political and legal contestation in the post-9/11 period. European civil society and parliamentary attempts to build a constraining consensus on these uses of force were in part predicated on the assumption that the security emergency that begat these military actions would decrease and that a return to the *status quo ante* might be achieved. More than two decades after the first use of armed drones, however, no such prospect of a decline in their use seems likely. Instead, armed drone use by the US and the UK governments has become institutionalised and with this the earlier intensity of the opposition to their controversial use has subsided. Indeed, far from armed drone use being restricted as an exceptional form of warfare, the opposite has become the case. As a result, the legality of using armed drones for non-battlefield targeted killing will remain both politically and legally contested and there will be no willingness to reconcile the divisions surrounding what is a divisive issue in contemporary international society. What this means in terms of a European consensus is that if one or more EU or NATO states engage in the use of armed drones for non-battlefield targeted killing and justify this in terms of a broadened conception of imminence, it can be expected that the previous practice of publicly acquiescent behaviour on the part of European governments, especially the French and German ones, will continue.

Conclusion

What this Element has demonstrated is the diverse responses in both military policy and legal strategising amongst the three leading European states to the changing threat environment presented by non-state terrorist actors in the post 9/11 world and the opportunities to respond to these in new and different ways presented by the technological innovation of armed drones. The United States responded to this threat with an innovating strategy in both policy – non-battlefield targeted killing through the use of armed drones and in law, a broadened interpretation of imminence. Subsequently, the United Kingdom followed this pathway in both practice and in its legal strategising. Yet the technological push of new drone technologies was not enough to persuade the

French and German governments to innovate in the same way and follow the UK government and embrace a broader conception of imminence.

Why this has been the case sets up the overarching puzzle that has motivated the Element. In this final section, we show how the four variables that we identified in the Introduction as explaining the differences in legal strategising have operated across the three cases. The four variables are (i) domestic politics; (ii) individual leadership; (iii) alliance politics; and (iv) precedent-setting. Before we turn to this, the first two parts of the Conclusion focus on (i) the dynamic and changing threat environment which we argue makes the issue of non-battlefield targeted killing by armed drones one that will remain salient in the years ahead, especially given the repercussions that might follow the US withdrawal from Afghanistan and (ii) the enabling possibilities for the use of force and associated legal strategising opened up by the development and proliferation of armed drone technologies. The final part of the Conclusion reflects on the dynamic interplay between enabling and constraining conceptions of the law, the importance of legal justification in international society, and the dangers of treating legal rules as if they were endlessly manipulable. We argue that instrumentalising the law in this way risks weakening the constraints on the use of force in contemporary international society.

The Dynamic and Changing Threat Environment

A key enduring theme of this Element has been the dynamic nature of the debate over the legality of the use of force for counterterrorism purposes in the post-9/11 period among the major European military states. An important driver of that dynamism has been the evolving and changing nature of the international security environment in which the use of force is contemplated. A key assumption of successive US administrations was that the use of force by drones and other military means would degrade the threat from violent non-state terrorist groups such as ISIL in Iraq and Syria. The threat from the latter has largely diminished, but this threat has not gone away. Indeed, what re-emerged in 2021 was a potential state-level threat with the Taliban's restoration to power. The US withdrawal from Afghanistan in the summer of 2021 demonstrated both that the US government's effort to eradicate the safe havens for terrorists was overambitious and that Washington and its allies have to accommodate their strategy to dealing with the potential once again of Afghanistan becoming the host to terrorist training camps and a space for the active instigation of international terrorism.

The end of the US military presence in Afghanistan should not be seen as indicating an end to the willingness of the US government to engage in remote warfare. Nor is it likely that this will result in an end to drone strikes. It is no accident that the unique combination of capabilities offered by armed drones favours the West's chosen approach to militarised counterterrorism. Characterised by a preference for light-footprint remote interventions and driven by an impulse for risk reduction, remote warfare is a strategy adopted in response to the choice of addressing terrorism through military means. The changing threat environment with regard to counterterrorism, post 9/11, was a major driver of change in the ways that states responded to the international terrorist threat. In this new context, states have developed additional military capabilities to enable them to meet that threat.

New Capabilities, New Temptations

Armed drone technology presents decision-makers with a precision military instrument that provides new opportunities for using force in a counterterrorism role. The reach of the armed drone presents states with the capability to execute strikes against targets which were previously out of reach for technical and political reasons. Drones uniquely allow the opportunity to respond in real time to actionable intelligence. How far, then, has this new capability lowered the threshold to the use of force? Based on the case studies in this Element, the following assessments can be made. There is a strong argument to be made that had the UK government not possessed the Reaper drones, it would not have undertaken the 2015 strike against UK nationals in Syria. There are three reasons for this: first, the drone's unique ISR capabilities may have been decisive in the identification of Khan and the other ISIL associates as targets with the opportunity to strike them in real time. Second, the drone's capacity for persistent air surveillance ensured that the strike could take place when there was the lowest possible risk of civilian casualties. Third, while a strike was technically possible with crewed aircraft (Tornado GR-4s or Typhoons) or Special Forces, the calculation of risk due to the possibility of casualties to UK military personnel and the political costs of failure would have significantly increased the political risks of such an attack, exercising a strong inhibiting effect on decision-makers.

By contrast, the French government's commitment to staying within a conventionalist legal framework has so far proved a stronger imperative than the technological possibilities opened up by armed drone technology. Even though the French government acquired Reaper drones, their initial use was limited to the ISR role in counterterrorist missions. Only after the Paris attacks in November 2015 did the French move to acquire an armed strike capability for their drone force. As soon as this became operational, armed strikes were conducted in the Sahel. Importantly, however, there has been to date no equivalent of the UK government's non-battlefield targeted killing by Paris. As we argued in Section 2, should Paris decide to use force for non-battlefield targeted killing, policymakers are most likely to invoke an exceptionalist moral justification, mindful of the risks of eroding the rules-based order, the upholding of which is seen across the domestic political spectrum as essential to French security.

For the German government, there has been more than a decade-long debate regarding the acceptable use of armed drones for the German military. Despite the call from German soldiers deployed in Afghanistan for armed drones to provide force protection, the association of this technology with US and UK non-battlefield targeted killing has generated a deep reluctance on the part of successive German governments to acquire this capability. Ironically, only once German participation in the NATO mission in Afghanistan had ended, did the German government, in the form of the 2021 Coalition Agreement, commit to acquire armed drones for the sole purpose of force protection. Despite reservations across the political spectrum about armed drone acquisition leading to US-type non-battlefield targeted killing, the unique capabilities of the drone have proved too tempting to resist for the German military and political leadership. The question for the future is whether having crossed the Rubicon of armed drone acquisition, new roles and missions will be found for these capabilities.

Recent evidence from the 2020 Nagorno-Karabakh conflict, the Ethiopian civil war in 2021, and the war in Ukraine demonstrates the immense potential of drones. In the case of the Azerbaijan-Armenia conflict, MALE Bayraktar TB2s and loitering munitions such as the Israeli Harop and Harpy played a major role in the stunning near-total victory by Azerbaijan. It was also the new element of affordable airpower in the form of drones that was decisive in the success of the Ethiopian government's counterterrorist strategy.[115] In the case of the Ukraine–Russia war, drones have operated in both the ISR and combat roles, providing key intelligence and targeting information but also as weapons platforms used extensively by both sides.[116]

The speed with which drones are proving their worth in modern conflicts suggests that the clamour to incorporate them into mainstream military arsenals as a necessary part of national defence capabilities may have implications for

---

[115] Zwijnenburg, W., 'Turkish Drones Join Ethiopia's war, Satellite Imagery Confirms', 11 January 2022, https://paxforpeace.nl/news/blogs/turkish-drones-join-ethiopias-war-satellite-imagery-confirms.

[116] Kramer, A., 'We Heard It, We Saw It, Then We Opened Fire', *The New York Times,* 23 October 2022, www.nytimes.com/2022/10/23/world/europe/ukraine-russia-drones-iran.html.

future non-battlefield targeted killing. As armed drones, in whatever form, become a core part of military inventories, governments will find themselves with an inexpensive and versatile instrument that can readily be employed to degrade and destroy insurgent groups that they perceive as a threat to national security.

The diffusion of armed drone technology to an increasing number of states beyond the Western alliance, and the expansion in the use of force this makes possible, suggests that policymakers and their legal advisors can be expected to reach for a broadened interpretation of imminence to justify counterterrorist uses of force in a situation where that state has not consented to the use of force and there is no UN Security Council authorisation for this.

Explaining the Differences in Legal Strategising

In answering the puzzle of how the three major European military states came to diverge so significantly both in their responses to the new US use of force post 9/11, and in their own individual relationship to the advent of armed drones, the previous sections have demonstrated the operation of four key interrelated variables.

Domestic Politics

For the UK government, the domestic political context played a key role in their innovating legal strategising. This was due in large part to the heightened sensitivity to the use of force in parliament following the contestation about the post-9/11 wars and the invasion of Iraq in particular. The fact that the UK Parliament had passed a resolution limiting the authorisation for the use of force against terrorist targets in Syria pushed Cameron to publicly espouse a broadened conception of imminence to justify the killing of Khan. Here, Cameron was leveraging his constitutional authority as prime minister under the Royal Prerogative which allows the latitude to use force when there is no time to consult parliament.

Where the United Kingdom's domestic constitutional context enabled Cameron's use of force, the German constitution (the Basic Law) imposes highly restrictive constraints on the use of force that cannot, as in the British case, be circumvented by executive discretion. The German constitution is also a product of the wider, post-Second World War cultural reluctance to endorse any use of force unless this is directly related to the self-defence of the German state. This is itself a product of the history and trauma of Germany in the twentieth century. Parliamentarians are deeply responsive to these sensitivities and concerns, and this is reflected in the fact that there is no support for non-battlefield targeted killing in the German body politic. It was with great reluctance, and belatedly, that the 2022 Coalition government decided to acquire drones with the capacity to use them in an armed role. But this has been agreed on the explicit understanding that unlike the UK and French governments, their armed role will be limited to force protection purposes only.

By contrast, domestic politics in France has played neither an enabling (United Kingdom) nor a constraining (Germany) role in the use of force by armed drones. The nature of the Fifth Republic means that the constitutional position of the French President limits the scope for domestic political debate about defence and foreign policy. In turn, the framing of the Fifth Republic also reflects French history and political culture as a state with a tradition of using force in support of its interests and values internationally, but also in upholding its responsibilities as a permanent member of the UN Security Council. The consequence of this domestic context means that the dominance of the Presidency in the area of foreign and security policy has the effect of limiting both parliamentary and civil society debate in France. If as Lushenko et al. argue (2022b; see also Lushenko, 2022), public support for the use of force in France is strongly correlated with 'multilateralism', then French policymakers could face difficult trade-offs if they were to engage in non-battlefield targeted killing in Mali and elsewhere that lacked state consent or in the absence of this, UN Security Council authorisation. As we discussed in Section 2 in relation to Margot-Mahdavi's argument, the French government could engage in future non-battlefield targeted killings, but offer no legal justification – not even the moral exceptionalist one suggested by Vilmer that drew on the Kosovo case. Such a strategy of 'silence' (Margot-Mahdavi, 2020) would lack sustainability at both home and abroad if it continued for a long period, alongside an extensive practice of highly visible non-battlefield military air strikes, given France's declaratory commitment to a rules-based legal order.

Individual Leadership

Leadership is the second key variable in explaining the differences in legal strategising and manifests itself differently in each of the states analysed. A central factor in the UK's innovating legal strategising was the individual leadership of Prime Minister David Cameron. This was a critical factor in explaining the UK's decision to use armed force, including armed drones, against terrorist targets in Syria. Cameron was the first prime minister to create the role of National Security Advisor and the body of the National Security Council. Mindful of the importance of ensuring that he had robust legal justifications to defend his use of force, he also was the first prime minister to appoint a legal advisor in the Prime Minister's Office. The latter was an explicit attempt by Cameron to have immediate counsel about what was legally possible internationally in order to maximise his freedom of action. As is evident from his speeches justifying the UK's use of force against ISIL and his reflections in his memoir, he was morally seized by the need to act against individual members of ISIL who were perceived as posing a direct and immediate threat to the United Kingdom. Although his successors may have faced different circumstances and less immediate threats of this kind, it is noteworthy that none of his four successors as prime minister at the time of writing have engaged in any further non-battlefield targeted killings. The element of individual leadership was less prominent as a consideration under Chancellor Angela Merkel's leadership of several German coalition governments. However, it became a significant factor after the 2022 Coalition Agreement was signed when the new government took the decision on legal grounds to withdraw from any military involvement in Syria. The agency in question here was the new German Foreign Minister Annalena Baerbock. She explained that her party's opposition was not in relation 'to the mission's goals, but our reservations about its legal basis at that time'.[117] The Coalition government's decision to end their participation in the military operation in Syria because it lacked a firm 'legal basis' shows how far a constraining view of the law shaped the new foreign minister and Chancellor's decisions. This was in contrast to the interpretation of the law in relation to Germany's military participation in Syria that had been proffered by the previous Chancellor and her ministers. Here, Baerbock was leveraging her position as a leader within the new coalition to achieve her foreign policy beliefs and preferences.

Individual leadership is not a significant factor in explaining the differences between French legal strategising on the one hand, and UK and German legal strategising on the other. This is because successive occupants of the Elysee Palace have acted with continuity in regard to the French use of force in the Sahel and elsewhere. Although the French government has engaged extensively in the use of force for counterterrorism purposes in Mali and the wider Sahel, including with the use of armed drones, they have to date done so on the legal conventionalist basis of a mix of host state consent, collective self-defence, and UN Security Council authorisation.

Alliance Politics

Turning to the third variable of alliance politics in explaining the differences in legal strategising, the UK and US governments' embrace of the broadened conception of imminence grew out of their close transatlantic partnership.

---

[117] 'Speech by Foreign Minister Annalena Baerbock in the German Bundestag on the extension of the mandate to counter IS', Reuters, 14 January 2022, www.auswaertiges-amt.de/en/newsroom/news/anti-is-mandate/2506642.

UK legal strategising was not merely a replication of its US partner; instead, the new US–UK interpretation of the self-defence rule appears to have been as much a product of UK official legal thinking – the 'Bethlehem principles' – as it was the 'lawyering' of the Obama administration. The Wright speech in 2017 was the fullest expression of the government's approach to imminence, providing a post hoc intellectual justification for the killing of Khan in 2015. It also can be read, particularly in its timing, in the context of the EU Referendum vote and the election of Donald Trump in 2016. The Brexit decision heightened the desire of the United Kingdom to create a closer relationship with Washington as part of its 'Global Britain' ambition, and the inauguration of Trump provided the stimulus for London to articulate a legal position anticipating a more permissive attitude to counterterrorism on the part of the incoming Trump administration. Unlike France and Germany, the UK government, post Brexit, was seeking a post-European focus to its foreign policy and championing the same legal justification as the United States provided an opportunity to more closely align itself with Washington. As an associated aspect of that post-European approach, the United Kingdom sought to solicit wider acceptance for a broadened conception of imminence. This diplomacy led to the public embrace of the Bethlehem principles by the Australian government set out in Brandis's 2017 address. The Australian government shared the UK government's commitment to strengthen their security relationship post Brexit.

If alliance considerations supported the UK's legal innovating strategy, a contrasting set of impulses led the German government to avoid openly criticising the new legal justifications mobilised by their transatlantic partners. In the German case, alliance considerations manifested themselves in Germany avoiding taking public positions which signalled its disapproval of the legal approaches adopted by its major NATO allies. Despite the deep unease with non-battlefield targeted killing on the part of successive German governments, Berlin has been very careful not to openly criticise the legal justifications mobilised by the US and UK governments. Whereas for the United Kingdom, the embrace of the broadened doctrine of imminence was a way of deepening US–UK relations, for Germany, this move by its close allies represented an awkward development that was in conflict with both the public and political mood of the country. This tension has been most clearly on display in relation to the legal dispute over German governments allowing the United States to use the Ramstein Air Base to support US non-battlefield targeted killing as part of its wider war on terror. The legal dispute over whether the German government has been properly exercising its responsibilities in relation to Ramstein remains the subject of dispute in the courts. Underlying the government's claim that it is living up to these responsibilities is the determination not to jeopardise the wider alliance relationship with the United States and, in so doing, damage in the words of the 2015 Cologne Court Judgement, the 'foreign and defense policy interests' of Germany. As a result, the German government finds itself in an uncomfortable position and the way it navigates this is to engage in a legal strategising of silence in response to the state practice and *opino juris* of its closest allies.

For France, like Germany, alliance considerations were at play, but resulted in a muted response. Hitherto, French governments have not provided any expressions of support for the US and UK government's legal justifications in support of a broadened conception of imminence in relation to UK and US non-battlefield targeted killings. At the same time, neither have they criticised or condemned these actions. By remaining silent in the face of US and UK legal justifications for their non-battlefield targeted killings, the French government like the German one is arguably acquiescing in legal claims made by the US and UK governments rather than explicitly challenging the US and UK claims for a more expansive interpretation of self-defence. Alliance politics are clearly at work here in shaping French legal strategising in relation to its allies. As in the German case, this silence with regard to the actions and legal justifications of the allies is motivated by concerns that publicly articulating such contested legal issues would lead to a further deterioration in relations with London and potentially antagonise Washington. French silence is also influenced by its desire not to take a position at odds with Germany in relation to the two 'Anglo-Saxon' powers. Wherever possible, France seeks a unified position within the EU, especially in relation to Germany.

Precedent-Setting

Different perceptions of how far the precedents set by the US and UK governments might erode existing legal restraints on the use of force shape the differences in legal strategising between the UK government and its continental neighbours. For the UK government, its legal innovation was believed to be necessary, proportionate, and the precedent set controllable, whereas for France and Germany, this action and associated legal rationale risked emboldening others in the future. Attempts to both invoke a new legal precedent and control it at the same time were evident in Wright's 2017 speech. The Attorney General emphasised that the UK government was not 'dispens[ing] with the concept of imminence', but the meaning that he and the UK government now gave to it was in effect advancing new law in classic legal innovating fashion. At the same time, Wright went out of his way to distance this act of UK law-making from a 'doctrine of pre-emptive strikes against threats that are more remote', recognising that to extend the broadened notion of imminence too far could 'diminish the importance of a rules-based international order'. However, such an act of legal strategising could not escape the fact that the UK government through Wright's lecture was inviting other states to use force on the same legal basis – the broadened conception of imminence – as Cameron had used to justify the killing of Khan.

The silence of successive German governments has reflected a determination not to support the US and UK claims for a broadened conception of imminence. This stems in part from the alliance considerations discussed above, but also from domestic considerations. There is a strong consensus among members of the German parliament, especially on the left, that the use of armed drones risks lowering the threshold to the use of force and the first step on a slippery slope towards the adoption of autonomous 'killer robots'.[118] Due to these considerations the German government has found itself conflicted between on the one hand not criticising its US ally while on the other, avoiding any parliamentary backlash by being seen to lend legal support to the claims made by the UK and US governments.

The differences between German legal strategising, on the one hand, and UK legal strategising on the other go wider than intra-European considerations of not rocking the boat with its European allies. Berlin has had to weigh alliance considerations against its concern that the US and UK governments are extending the right of self-defence in ways that might set dangerous precedents that erode existing barriers to the use of force. Unlike the situation over the invasion of Iraq in 2003, however, these US and UK legal innovations have not been judged to date to be serious enough to warrant stoking alliance tensions with the US and UK governments. Their preferred response has been silence.

French governments have shared the German concern about legally endorsing the new US and UK uses of force because of the dangerous precedent this might set. However, Paris has also remained silent, in contrast to the position it took in 2003 over the Iraq War, in response to the new legal justifications mobilised by the US and UK governments. French governments have not considered it worthwhile to openly challenge London and Washington on this issue in part due to their P5 membership of the UN Security Council. As a permanent member, the French government has a structural incentive to be seen to be upholding the rules-based international order and it has interpreted this as requiring a more restrictive interpretation of the legal rules governing the recourse to force. At the same time, this restrictive interpretation might not be

---

[118] See Posener, A., 'The Drone a Chance', *De Zeit*, 21 October 2020, www.zeit.de/politik/ deutschland/2020-10/bundeswehr-bewaffnete-drohnen-toeten-hemmschwelle.

sustainable in Mali or elsewhere in the future. Faced with such a potential eventuality, French policymakers can be expected to pursue a path that they believe will have the least likelihood of setting precedents that others might emulate in ways that both damage French interests and weaken the restraints on the use of force internationally. Having shown how the operation of the four variables explains the differences in legal strategising between the UK, French, and German governments, the final part of the Conclusion turns to the place of legal strategising in the future international order.

Legal Strategising – Implications for the International Order

What is at stake in the debate over precedent-setting is weighing up how far other governments would emulate the legal justifications used by the UK and US governments to legitimate uses of force as against the dangers to national security of inaction in the face of terrorist threats. As Wright and Vilmer appreciated, this means controlling the use of precedents to enable military action by yourself and your allies while constraining the use of force on the part of others. But the problem with Vilmer's moral exceptionalist approach, and even more so Wright and the UK government's legal innovating one, is that precedents are rarely under the control of those who set them.[119]

Despite the UK government's attempt to limit the circumstances under which states could legitimately invoke the 'Bethlehem principles', the Turkish government's use of force for counterterrorist purposes in Syria in October 2019 invoked the language of imminence as part of its claim to be acting pursuant to Article 51 of the Charter. In a letter dated 9 October 2019 to the UN Security Council, Turkey's Permanent Representative informed it that Turkey had used force in Syria 'in line with the right of self-defence as outlined in Article 51 of the Charter of the United Nations, to counter the imminent terrorist threat'.[120] President Erdogan stated that 'some countries eliminate terrorists whom they consider as a threat to their national security, wherever they are. Therefore, this means those countries accept that Turkey has the same right'.[121] Erdogan was referring here to the permissive interpretation of the self-defence rule employed by the US and UK governments, with the Turkish Ambassador using the legal

---

[119] A pioneering discussion of how precedents set by one government can be used by others is Franck T.M and Weisband E. 1972. *Word Politics: Verbal Strategy Among the Superpowers.* Oxford University Press: New York. *Word Politics.*

[120] Letter dated 9 October 2019 from the Permanent Representative of Turkey to the United Nations addressed to the President of the Security Council, https://documents-dds-ny.un.org/doc/UNDOC/GEN/N19/309/32/PDF/N1930932.pdf?OpenElement.

[121] Quoted in Gibson, J., 'Opinion: We're quickly moving toward a world where drone executions are the norm', *Los Angeles Times*, 13 November, 2019, www.latimes.com/opinion/story/2019-11-13/drone-killings-war-syria-turkey.

language of an 'imminent terrorist threat' to locate Turkey's use of force within the precedents set by London and Washington. As leading international lawyer Oona Hathaway wryly observed, 'where could Erdogan possibly have gotten the idea that he could use a claim of self-defense to justify launching an aggressive attack on the sovereign territory of another state? He got it from us. Erdogan was exploiting a loophole the United States and other members of the Security Council have created and expanded over time' (Hathaway, 2019). Significantly, the Turkish example shows how one innovation can lead to another. Although citing precedent, Turkey was also expanding the notion of imminence that the United Kingdom had advanced, by moving beyond the idea of targeting individuals actively planning attacks, to the much broader category of 'terrorists' as Ankara would define them. In doing this, the scope of what counts as imminence is further being expanded to enable more permissive uses of force.

What Hathaway is articulating here is a concern shared across civil society organisations in the United States and Europe. It is also one that resonates with many European governments, but they feel unable for the political and legal reasons we have discussed throughout the Element to articulate these publicly. As a result, and as we set out at length in Section 4, the possibility of re-establishing the previous European consensus on what constitutes a legitimate right of self-defence is a distant one. It is this recognition that there is little possibility of achieving a new consensus that would constrain future uses of force that inhibits the European allies of the UK and US governments from criticising the new interpretations of imminence advanced by London and Washington.

Civil society groups and academic communities that are not constrained by these alliance considerations have attacked the new interpretation of the self-defence rule proffered by London and Washington as inviting other actors to do the same, bringing with it a dangerous erosion of the constraints on the use of force. In response to these criticisms, UK policymakers would respond that a limited expansion of the imminence doctrine is operationally necessary to meet the current threat and that this ought to be accepted as a reasonable contemporary legal interpretation of the Caroline standard. For them, the priority is the ability to meet the immediate threats from terrorism even if this sets new – albeit what they envisaged as limited – legal precedents for the preventive use of force. Moreover, the criticism that other states might be emboldened to use force in the future because of the precedents set by London, Washington, and Ankara could be countered by the rebuttal that others, especially the West's adversaries, would not hold back their own uses of force in times of crisis even if the UK and US governments continued to operate within a traditional legal conventionalist understanding of the Caroline standard.

Such a criticism opens up the counterfactual question as to how the transatlantic allies would have responded to other states leveraging the broadened conception of imminence for counterterrorist purposes in a context where they were all united in support of a narrow interpretation of imminence. Neither the UK nor US governments were constrained from using force for non-battlefield targeted killing because of the lack of a plausible legal justification, showing the enabling power of the law, especially when it comes to the use of force. But imagine if Russia or China, for example, had mobilised a broad interpretation of imminence in relation to the terrorist threat. Two possible pathways would open up here. The first is that in a context where the UK, French, German, and US governments were all committed to a narrow interpretation of imminence, such a legal justification would have lacked legitimation from these key states. The knowledge that such a practice would lack legitimation from these states and potentially wider international society would be unlikely to deter such actions if the stakes were high. To think otherwise is to expect too much from the constraining power of the law. However, actions that lack a strong degree of legitimation because of their perceived illegality may lead to adverse material consequences if states mobilise economic and political pressures against the government that is perceived to be in violation of the established rules. The extent to which states making controversial legal claims might be censured and sanctioned effectively by those seeking to maintain conventionalist legal interpretations will depend heavily on their own national power positions, geopolitical security considerations, and wider alliance affiliations and security guarantees.

The second outcome that might be envisaged where Russia and China were the legal innovating states is that far from condemning these, there is instead legal emulation on the part of some or all of the transatlantic grouping of states. However, in the balance between condemnation and emulation in relation to the US and UK government's actual practice 'of non-battlefield targeted killing', it is striking how so few states have emulated this. At the same time, there has been no condemnation of the US and UK governments from other Western-orientated states, with condemnation reserved for members of the legislature, human rights NGOs, and experts in the states engaging in non-battlefield targeted killings. The fact that it is the United States that has driven this new legal practice, supported later by the UK government's own legal innovation, and that their closest allies have acquiesced in this, supports the realist contention that international law is all too enabling of state power, especially when it comes to the use of force.

However, this Element has also shown that there is a counter-narrative to tell in relation to the constraining power of the law. As Bull and the English School have long argued, the rules are not 'infinitely malleable and do circumscribe the range of choice of states' (1977: 45). Policymakers seek to use the law when they want to constrain uses of force by themselves and others in pursuit of their normative positions. The real test of the constraining power of the law would be if legal advisors argued that a particular use of force was unlawful, and as a result, policymakers refrained from the action. Although there is no specific incident that we can point to, there is evidence from Cameron's memoir and other sources that the UK prime minister was frustrated by the legal advice that he received prior to August 2015 that he saw as limiting the UK's ability to strike at terrorist targets and engage more proactively in the humanitarian crisis in Syria. Cameron reflected in his memoir on the risks of policy being thwarted by the 'tendency to seek out legal obstacles' to the use of force. He went on, '(HISTORIANS note) we need legal advice. But we should consider what might work and then ask lawyers – not the other way round' (Cameron, 2019: 452). When it came to asking Jeremy Wright for 'legal advice' – or legal cover as we would argue – he had an Attorney General who was willing to interpret the existing legal rules in a manner that made possible a use of force that a more constrained interpretation of those rules would have ruled out.

The core thesis of this Element is what counts as lawful action is contested through argumentation over the relevant rules and their application. International society, Bull famously argued, depends on the 'settled expectations that states have about one another's behaviour' (Bull, 1977: 45). The UK government acknowledges the importance of not jeopardising the agreed legal rules on self-defence as can be seen in Wight's IISS address. However, this commitment rests uneasily with its innovating legal strategising in relation to the use of armed drones. The problem is that the UK and US's willingness to engage in legal strategising in relation to the use of armed drones, and earlier in relation to the use of force against Iraq, is perceived by many other states as stretching the boundaries of legitimate legal interpretation beyond breaking point. This invites others to mount a broader assault on international law itself as both fragile and hypocritical in the hands of the powerful. This can arguably be seen in Putin's willingness to advance legal justifications for his actions in Ukraine that fail to meet 'the laugh test' (Franck, 2006: 96) of plausible legal justification. Yet in the face of a Russian use of force that clearly fails 'the laugh test', what is striking and concerning is that many states in the global south, including Brazil, China, India, and South Africa, have chosen not only to refrain from condemning Russia's invasion of Ukraine, but also not to dismiss his derisive legal defence of the annexation of Ukrainian sovereign territory that blatantly challenges the 'settled expectations' of legal justification in international society. The absence of moral and political censure from large parts of the Global South has allowed Putin to level charges of hypocrisy at Western states who have been united in condemning Russia's flagrant breach of UN Charter principles.

It would be an unfair exaggeration to blame the legal innovations undertaken by the US and UK governments since 2001 and 2015 respectively for Russia's blatant contempt of international law in 2022–23. Nevertheless, in any account of the perceived fragility and hypocrisy of international law, the United States, United Kingdom, and Australian tortured legal justifications for removing Saddam Hussein from power in Iraq must loom large. Expanding the boundaries of self-defence through the 'Bethlehem principles' has enabled new uses of force – including through the use of armed drones – that has further contributed to the perception among many states that law is the handmaiden of power. The worry here is that too many governments start believing that the rules are 'infinitely malleable', thereby invalidating Bull's cardinal test for the existence of an international society and the order this makes possible. *Drones, Force and Law* has shown the importance of policymakers, especially in the most powerful states, recognising that they have to pass the 'laugh test' of legal justification when using force. Otherwise, there is the risk that the use of force expands still further in a context where the proliferation of drones and other lethal technologies place in the hands of policymakers tools for the prosecution of still more violence in international society.

Abbreviations

AFISMA – African-led International Support Mission to Mali
AJIL – American Journal of International Law
APPG – All Parliamentary Party Group
AQ – Al-Qaeda
AQAP – Al-Qaeda in the Arabian Peninsula
AUKUS – Australia-United Kingdom-United States
AUMF – Authorization for Use of Military Force
CDU – Christian Democratic Union of Germany
EC – European Council
ECCHR – European Centre for Constitutional and Human Rights
EXComm – Executive Committee
FCO – Foreign and Commonwealth Office
FDP – Free Democratic Party
GWOT – Global War on Terrorism
ICJ – International Court of Justice
IHL – International Humanitarian Law
IICK – Independent International Commission on Kosovo
IISS – International Institute for Strategic Studies
IRSEM – Institute for Strategic Research at the Military School
ISIL – Islamic State in the Levant
ISR – Intelligence, Surveillance and Reconnaissance
JCHR – Joint Committee on Human Rights
LAWS – Lethal Autonomous Weapon Systems
MINUSMA – The United Nations Multidimensional Integrated Stabilization
    Mission in Mali
MUJWA – Movement for Oneness and Jihad in West Africa
NIAC – Non-International Armed Conflict
P5 – Permanent 5 Members of the UN Security Council
SDP – Social Democratic Party
UKNSC – United Kingdom National Security Council
UNSCR – United Nations Security Council Resolution
WMD – Weapons of Mass Destruction

References

Ahmed, Y. 2018. RWUK taking UK government to court over refusal to disclose legal basis for targeted killings. *Rights & Security International*, 20 July. www.rightsandsecurity.org/impact/entry/rights-watch-uk-taking-uk-government-to-court-over-refusal-to-disclose-legal-basis-for-targeted-killings.

Akande, D. & Milanovic, M. 2015. The constructive ambiguity of the security council's ISIS resolutions. *EJIL: Talk!* 21 November. www.ejiltalk.org/the-constructive-ambiguity-of-the-security-councils-isis-resolution/.

Amaral, N. & Dorsey, J. 2021. *Military Drones in Europe*. Chatham House.

Banka, A. & Quinn, A. 2018. Killing norms softly: US targeted killing, quasi-secrecy and the assassination ban. *Security Studies*, 27(4), 665–703.

Bentley, M. 2014. *Weapons of Mass Destruction in US Foreign Policy*. Routledge: London.

Bethlehem, D. 2012. Self-defense against an imminent or actual armed attack by nonstate actors. *American Journal of International Law*, 106(4), 770–7.

Betts, R. 2003. Striking first: A history of thankfully lost opportunities. *Ethics & International Affairs*, 17(1), 17–24.

Bhuta, N. 2015. On preventative killing. *EJIL: Talk!* 17 September 2022. www.ejiltalk.org/author/negal-bhuta.

Birdsall, A. 2022. New technologies and legal justification: The United Kingdom's use of drones in self-defence. *Global Constitutionalism*, 11(2), 197–216.

Boyle, M. J. 2020. *The Drone Age: How Drone Technology Will Change War and Peace*. Oxford University Press: New York.

Brunnée, J. & Toope, S. J. 2011. Interactional international law: An introduction. *International Theory*, 3(2), 307–18.

Brunstetter, D. R. & Ferey, A. 2021. Armed drones and sovereignty: The arc of strategic sovereign possibilities. In Lushenko, P., Bose, S., & Maley, W. (eds.), *Drones and Global Order: Implications of Remote Warfare for International Society*. Routledge: London.

Bull, H. 1977. *The Anarchical Society: A Study of Order in World Politics*. Macmillan: London.

Buzan, B. 2015. The English School: A neglected approach to international security studies. *Security Dialogue*, 46(2), 126–43.

Buzan, B. & Schouenborg, L. 2018. *Global International Society: A New Framework for Analysis*. Cambridge University Press: Cambridge.

Byers, M. 1999. *Custom, Power and the Power of Rules: International Relations and Customary International Law.* Cambridge University Press: Cambridge.

Byers, M. 2003. Preemptive self-defense: Hegemony, equality and strategies of legal change. *The Journal of Political Philosophy,* 11(2), 171–90.

Byers, M. 2021. Still agreeing to disagree: International security and constructive ambiguity. *Journal on the Use of Force and International Law,* 8(1), 91–114.

Cameron, D. 2019. *For the Record.* HarperCollins. Kindle Edition.

Carr, E. H. 1939. *The Twenty Years' Crisis, 1919–1939: An Introduction to the Study of International Relations.* Macmillan: London.

Charbonneau, C. 2017. Whose 'West Africa'? The regional dynamics of peace and security. *Journal of Contemporary African Studies,* 35(4), 407–14.

Chayes, A. 1974. *The Cuban Missile Crisis: International Crisis and the Role of Law.* Oxford University Press: New York.

Chayes, A. H. & Chayes, A. 1998. *The New Sovereignty: Compliance with International Regulatory Agreements.* Harvard University Press: Cambridge, MA.

Chrisafis, A. 2015. France launches first airstrikes against ISIS in Syria. *The Guardian.* www.theguardian.com/world/2015/sep/27/france-launches-first-airstrikes-isis-syria.

Clark, I., 2005. *Legitimacy in International Society.* Oxford University Press: Oxford.

Clark, I. 2007. *International Legitimacy and World Society.* Oxford University Press: Oxford.

Clark, I. 2011. *Hegemony in International Society.* Oxford University Press: Oxford.

Clark, I. & Reus-Smit, C. 2007. Resolving international crises of legitimacy. *International Politics,* 44(2–3), 153–335.

Claude, I. 1966. Collective legitimization as a political function of the United Nations. *International Organization,* 20(3), 367–79.

Coker, C. 2001. *Humane Warfare.* Routledge: London.

Cvijic, S., Klingenberg, L., Goxho, D., & E. Knight. 2019. *Armed Drones in Europe.* Open Society European Policy Unit. https://opensocietyfoundations.org/uploads/2ded5bae-143a-45ee-9fc9-7d30dbf3b62d/armed-drones-in-europe-20191104.pdf.

Dorsey, J. & Paulussen, C. 2015. Towards a European position on armed drones and targeted killing: Surveying EU counterterrorism perspectives. *ICCT.* https://icct.nl/app/uploads/2020/10/ICCT-Dorsey-Paulussen-Towards-A-European-Position-On-Armed-Drones-And-Targeted-Killing-Surveying-EU-Counterterrorism-Perspectives.pdf.

Dunlap, C. 2001. Law and military interventions: Preserving humanitarian values in 21st conflicts. Prepared for the *Humanitarian Challenges in Military Intervention Conference Carr Center for Human Rights Policy Kennedy School of Government*. Harvard University: Washington, DC, 29 November. https://isc.independent.gov.uk/wp-content/uploads/2021/01/20170426_press_release_on_UK_Lethal_Drone_Strikes_in_Syria.pdf.

Dworkin, A. 2015. The EU and armed drones: Epilogue. *Global Affairs*, 1(3), 293–6.

Franck, T. M. 1990. *The Power of Legitimacy among Nations*. Oxford University Press: New York.

Franck, T. M. 2002. *Recourse to Force: State Action against Threats and Armed Attacks*. Cambridge University Press: Cambridge.

Franck, T. M. 2006. The power of legitimacy and the legitimacy of power. *American Journal of International Law*, 100(1), 88–106.

Franck, T. M. & Rodley, N. S. 1973. After Bangladesh: The law of humanitarian intervention by military force. *American Journal of International Law*, 67(2), 275–305.

Franck, T. M. & Weisband, E. 1972. *Word Politics: Verbal Strategy among the Superpowers*. Oxford University Press: New York.

Franke, U. 2016. *Proliferated Drones: A Perspective on Germany*. Centre for a New American Security. https://drones.cnas.org/wp-content/uploads/2016/05/A-Perspective-on-Germany-Proliferated-Drones.pdf.

Franke, U. 2017. *The Unmanned Revolution: How Drones Are Revolutionising Warfare*. D Phil, University of Oxford. https://ora.ox.ac.uk/objects/uuid:ab40b722-2613-478a-912c-5b06307a3435.

Green, J. A. 2009. *The International Court of Justice and Self-defence in International Law*. Bloomsbury: London.

Greenstock, J. 2016. *Iraq: The Cost of War*. Arrow Books: London.

Gregory, T. 2017. Civilian casualties, non-combatant immunity and the politics of killing: Review essay. *Critical Studies on Terrorism*, 10(1), 187–96.

Gusterson, H. 2016. *Drone: Remote Control Warfare*. MIT Press: Cambridge, MA.

Hathaway, O. 2019. Turkey is violating international law: It took lessons from the U.S. *The Washington Post*, 22 October. www.washingtonpost.com/outlook/2019/10/22/turkey-is-violating-international-law-it-took-lessons-us/.

Hathaway, O. A. & Shapiro, S. J. 2018. *The Internationalists: And Their Plan to Outlaw War*. Penguin Books: London.

Heller, J. K. 2019. The earliest invocation of the unwilling or unable. *OpinoJuris*, 19 March. http://opiniojuris.org/2019/03/19/the-earliest-invocation-of-unwilling-or-unable/.

Hendrickson, D. C. 2002. Toward universal empire: The dangerous quest for absolute security. *World Policy Journal*, 19(3), 1–10.

Henkin, L. 1979. *How Nations Behave: Law and Foreign Policy*. Columbia University Press: New York.

Holland, J. 2020. *Selling War and Peace: Syria and the Anglosphere*. Cambridge University Press: Cambridge.

Hurd, I. 2017a. *How to Do Things with International Law*. Princeton University Press: Princeton.

Hurd, I. 2017b. The permissive power of the ban on war. *European Journal of International Security*, 2(1), 1–18.

Hurrell, A. 2007. *On Global Order: Power, Values and the Constitution of International Society*. Oxford University Press: Oxford.

Jackson, R. H. 2000. *The Global Covenant: Human Conduct in a World of States*. Oxford University Press: Oxford.

Jennings, R. Y. 1938. The Caroline and McLeod cases. *American Journal of International Law*, 32(1), 82–99.

Johnstone, I. 2003. Security council deliberations: The power of the better argument. *European Journal of International Law*, 14(3), 437–80.

Kaag, J. & Kreps, S. E. 2014. *Drone Warfare*. Polity Press: Cambridge.

Kaempf, S. 2018. *Saving Soldiers or Civilians: Casualty-Aversion versus Civilian Protection in Asymmetric Conflicts*. Cambridge University Press: Cambridge.

Keating, V. C. 2022. Membership has its privileges: Targeted killing norms and the firewall of international society. *International Studies Quarterly*, 66(3). https://doi.org/10.1093/isq/sqac040.

Kelly, F. 2020. Sahel coalition: G5 and France agree new joint command, will prioritize fight against Islamic state. *The Defense Post*, 14 January. www .thedefensepost.com/2020/01/14/sahel-coalition-france-g5-islamic-state/.

Kitties, O. F. 2016. *Lawfare: Law as a Weapon of War*. Oxford University Press: Oxford.

Koskenniemi, M. 2002. *The Gentle Civilizer of Nations: The Rise and Fall of International Law 1870–1960*: Cambridge University Press: Cambridge.

Koskenniemi, M. 2005. International law in Europe: Between tradition and renewal. *The European Journal of International Law*, 16(1), 113–24.

Kreps, S. E. 2014. Flying under the radar: A study of public attitudes towards unmanned aerial vehicles. *Research & Politics*, 1(1), 1–7.

Kreps, S. E. 2016. *Drones: What Everyone Needs to Know*. Oxford University Press: Oxford.

Kreps, S. & Lushenko, P. 2023. Drones in modern war: Evolutionary, revolutionary, or both? *Defense and Security Analysis*, 15(3), 271–4.

Kreps, S. E. & Wallace, G. P. 2016. International law, military effectiveness, and public support for drone strikes. *Journal of Peace Research*, 53(6), 830–44.

Lobel, J. & Ratner, M. 1999. Bypassing the security council: Ambiguous authorizations to use force, cease-fires and the Iraqi inspection regime. *American Journal of International Law*, 93(1), 124–54.

Lushenko, P., Bose, S., & Maley, W. (eds.). 2022a. *Drones and Global Order: Implications of Remote Warfare for International Society*. Routledge: London.

Lushenko, P., Raman, S., & Kreps S. 2022b. Multilateralism and public support for drone strikes. *Research & Politics*, 9(2). https://doi.org/10.1177/2053168022 1093433.

Lynch, C. 2013. France's U.N. envoy: French military intervention in Mali is open ended. *Foreign Policy*, 12 January. https://foreignpolicy.com/2013/01/12/frances-u-n-envoy-french-military-intervention-in-mali-is-open-ended/.

MacDonald, J. 2017. *Enemies Known and Unknown: Targeted Killings in America's Transnational Wars*. Oxford University Press: Oxford.

Mayall, J. 2000. *World Politics: Progress and Its Limits*. Polity: Cambridge.

Mignot-Mahdavi, R. 2020. Le Silence des Agneaux: France's war against Jihadist groups. *ICCT Journal*, Research Paper, May 2020. https://icct.nl/publication/le-silence-des-agneaux-frances-war-against-jihadist-groups-and-associated-legal-rationale/.

Mignot-Mahdavi, R. 2023. *Drones and International Law: A Techno-Legal Machinery*. Cambridge University Press: Cambridge.

Milanovic, M. 2020. The Soleimani strike and self-defence against an imminent armed attack. *EJIL: Talk!* 7 January. www.ejiltalk.org/the-soleimani-strike-and-self-defence-against-an-imminent-armed-attack.

Mills, C. 2017. *ISIS/Daesh: The Military Response in Iraq and Syria*. House of Commons Library Briefing Paper, Number 06995, 8 March. https://research briefings.files.parliament.uk/documents/SN06995/SN06995.pdf.

Morgenthou, F, H. J. 1967. *Politics among Nations: The Struggle for Power and Peace*. Alfred A. Knopf: New York.

Moyn, S. 2021. *Humane: How the United States Abandoned Peace and Reinvented War*. Farrar, Straus, and Giroux: New York.

Murphy, S. D. 2005. The doctrine of preemptive self-defense. *Villanova Law Review*, 50(3), 699–748.

O'Connell, M. E. 2011. Remarks: The resort to drones under international law. *Denver Journal of International Law and Policy*, 39(4), 585–600.

O'Connell, M. E. 2020. The killing of Soleimani and international law. *EJIL: Talk!* 6 January. www.ejiltalk.org/the-killing-of-soleimani-and-international-law/.

Peevers, C. 2013. *The Politics of Justifying Force: The Suez Crisis, the Iraq War, and International Law.* Oxford University Press: Oxford.

Pompeo, M. 2020. *Secretary of State Pompeo Interviewed on Fox and Friends.* US Department of State. www.state.gov/secretary-michael-r-pompeo-with-steve-doocy-ainsley-earhardt-and-brian-kilmeade-of-fox-and-friends/.

Power, S. 2019. *The Education of an Idealist.* William Morrow: New York.

Recchia, S. & Thierry, T. 2020. French military actions in Africa: Reluctant multilateralism. *Journal of Strategic Studies,* 43(4), 473–81.

Reus-Smit, C. (ed.). 2004. *The Politics of International Law.* Cambridge University Press: Cambridge.

Rhodes, B. 2018. *The World as It Is: Inside the Obama White House.* Bodley Head: London.

Sands, P. 2005. *Lawless World: America and the Breaking of Global Rules.* Allen Lane: London.

Scheuerman, W. E. 1999. *Carl Schmitt: The End of Law.* Roman & Littlefield: Lanham, MD.

Schmidt, D. R. & Trenta, L. 2018. Changes in the law of self-defence? Drones, imminence, and international norm dynamics. *Journal on the Use of Force and International Law,* 5(2), 201–45.

Seldon, A. 2020. *May at 10.* Biteback. Kindle Edition.

Seldon, A. & Snowdon, P. 2015. *Cameron at 10: The Verdict.* William Collins: London.

Simpson, G. 2000. The situation on the international legal front: The power of rules and the rule of power. *European Journal of International Law,* 11(2), 439–64.

Simpson, G. 2004. *Great Powers and Outlaw States: Unequal Sovereigns in the International Legal Order.* Cambridge University Press: Cambridge.

Skinner, Q. 1974. Some problems in the analysis of political thought and action. *Political Theory,* 2(3), 277–303.

Skinner, Q. 1988. A reply to my critics. In Tully, J. (ed.), *Meaning and Context: Quentin Skinner and His Critics.* Polity Press: Cambridge, 231–89.

Stevenson, J. 1970. THE CAMBODIAN INCURSION: United States notifies U.N. Security Council. *International Legal Materials,* 9(4), 838–71.

The Bureau of Investigative Journalism. 2021. *Human Rights: Drone Warfare.* www.thebureauinvestigates.com/projects/drone-war.

The Independent International Commission on Kosovo. 2000. *The Kosovo Report: Conflict, International Response, Lessons Learnt.* https://reliefweb.int/sites/reliefweb.int/files/resources/6D26FF88119644CFC125698900 5CD392-thekosovoreport.pdf.

Trenta, L. 2018. The Obama administration's conceptual change: Imminence and the legitimation of targeted killings. *European Journal of International Security*, 3(1), 69–93.

Verdier, P. H. & Voeten, E. 2015. How does customary international law change? The case of state immunity. *International Studies Quarterly*, 59(2), 209–22.

Vilmer, J. B. J. 2015. When France arms its drones. *National Defence Review*, 96–101. www.jbjv.com/IMG/pdf/JBJV_2015_-_When_France_Arms_Its_Drones.pdf.

Vilmer, J. B. J. 2017. The French turn to armed drones. *War on the Rocks: Texas National Security Review*, 22 September. https://warontherocks.com/2017/09/the-french-turn-to-armed-drones/.

Vilmer, J. B. J. 2021. Not so remote drone warfare. *International Security*, 9 July.

Waddington, C. 2014. Understanding Operation Barkhane. *African Defence Review*, 1 August. www.africandefence.net/operation-barkhane-under-the-hood.

Wheeler, N. J. 2000. *Saving Strangers: Humanitarian Intervention in International Society*. Oxford University Press: Oxford.

Wheeler, N. J. 2003. The Bush doctrine: The dangers of American exceptionalism in a revolutionary age. *Asian Perspective*, 27(4), 183–216.

Wheeler, N. J. 2004. The Kosovo bombing campaign. In Reus-Smit, C. (ed.), *The Politics of International Law*. Cambridge University Press: Cambridge, 198–217.

Yoo, J. 2003. International law and the War in Iraq. *The American Journal of International Law*, 97(3), 563–76.

Yoo, J. 2004. Using force. *University of Chicago Law Review*, 71(3), 729–97.

Zehfuss, M. 2018. *War and the Politics of Ethics*. Oxford University Press: Oxford.

Acknowledgements

This Element is the latest product of over ten years of research on armed drones at the University of Birmingham (UoB). During this period, we have received external research funding from the Economic and Social Research Council, the Gerda-Henkel Foundation, and the Open Society Foundations (OSF), and we wish to express our thanks to all three bodies. We are also grateful for funding and supporting our work through the 2014 BPC, of which we were both Commissioners. The Commission concluded in its 2014 report that, 'The UK simply does not accept the specific US legal justification for using RPA for the targeted killing of AQ-related terrorist targets.' This position was one that was shared with the UK Government's major European allies and was one that was expected to carry on for the foreseeable future. The announcement by Prime Minister David Cameron in Parliament on 7 September 2015 that his government had authorised the killing of Reyaad Khan using an armed drone in August 2015 marked an abrupt break with the previous orthodoxy. The UK Government strongly aligned itself with the legal justifications being advanced by the Obama administration for the targeted killing of terrorists outside of recognised armed conflicts. This was the point of departure for the present study. As we wanted to conduct research into the possibilities of repairing this breach in the European position and establishing a new European legal and political consensus that could constrain the use of armed force, including armed drones, for counter-terrorist operations.

We have accumulated a large number of debts in conducting this research. We wish to thank those officials at the OSF who have supported the development of this work, especially Soheila Comninos. We have been greatly aided in our thinking by the contribution of an Advisory Group and wider study group meetings. In terms of the Advisory Group, we wish to thank the following: Jennifer Gibson, Professor Peter Gray, Chris Lincoln Jones, Sir David Omand, Elizabeth Quintana, and Professor Paul Schulte. In addition, we wish to thank Lord Hodgson and Lord Wallace of Tankerness for hosting two study group meetings in Parliament where earlier drafts of sections of the Element were workshopped. We wish to thank the then organisers of the APPG, Aditi Gupta and Camilla Molyneux, for organising these events and the following, all of whom attended one or both meetings: Sir Michael Aaronson, Nilza Amaral, Chris Cole, Anthony Dworkin, Jennifer Gibson, Harriet Hoffler, Emily Knowles, Dr Larry Lewis, Dr Jack McDonald, Elizabeth Minor, Professor Paul Schulte, and Anna de Courcy Wheeler. In addition, our thinking has been enhanced by conversations and interactions with the following: Yasmin Ahmad, Professor Dapo Akande, Rosalind Comryn, Neta Crawford, Adriana Edmeades, Delina Goxho, Dominic Grieve (former Attorney General), Professor Caroline Kennedy-Pipe, Dr Peter Lee, Harriet Moynihan, Professor Mary Ellen O'Connell, Jean-Baptiste Vilmer, Elizabeth Wilmshurst, Chris Woods, and Wim Zwijnenburg. We are particularly indebted to the Stimson Center and to Rachel Stohl in particular for giving us an opportunity to present our work and share our findings with the US and international drone community.

We are particularly indebted to those who have worked with us in developing the work on drones at UoB over the last decade. The research support provided by the following has greatly enriched this Element: Dr Lindsay Clark Maren Evensen, Dr Talat Farooq, Dr Jamie Johnson, Barney Jones, George May, Ellen Moser, Georgia Taylor, and Dr Christopher Wyatt. In addition, we have been expertly assisted on the administrative side by Guvinder Kaur Rajina and especially Dr Catherine Edwards. The latter deserves a special thanks for her unstinting professionalism and indeed patience as the authors struggled to balance conflicting pressures. Academic colleagues in the Department of Political Science and International Studies at Birmingham provided a fertile intellectual and supportive space to discuss ideas developed further in this Element. Here, we thank Professor Peter Gray, Dr George Kyris, Professor Scott Lucas, Dr Adam Quinn, Professor Mark Webber, and Professor Stefan Wolff.

The sensitive and secret nature of this topic makes research into it difficult and challenging. We have been fortunate to have the opportunity to interview some two dozen or so former military personnel, political and diplomatic officials, and members of the NGO community. Many of these individuals cannot be named and many of them have asked that their interviews remain strictly confidential. Nevertheless, we would note that access, especially to former and current legal advisors, has been challenging.

We want to thank the following for reading the manuscript in whole or in part: Dr Adriana Birdswell, Professor Michael Byers, Dr Jessica Dorsey, Dr Rebecca Mignot-Mahdavi, Professor Mark Saunders, Andreas Schüller, and especially Paul Lushenko for his detailed and incisive comments on the full manuscript. Our biggest thanks here goes to Professor Paul Schulte, a member of the BPC and Advisory Group, whose support, encouragement, and encyclopaedic knowledge of the subject have proved invaluable. He has gone above and beyond, reading drafts as the work has evolved. *Drones, Force and Law* has been written in conjunction with Jack Davies and Zeenat Sabur. Their background research, data collection and data analysis, and contribution to the writing of earlier drafts have greatly assisted us in the production of the final manuscript and we thank them for all their efforts.

Finally, we wish to offer our considerable thanks to the reviewers at Cambridge University Press and to Sarah Kreps for her important suggestions in how to develop the Element. We also thank Caroline Parkinson for all her support in the editing process of the Element, and especially John Haslam and Jon Pevehouse for their support of the project.

Cambridge Elements $^{\equiv}$

International Relations

Series Editors

Jon C. W. Pevehouse
University of Wisconsin–Madison

Jon C. W. Pevehouse is the Mary Herman Rubinstein Professor of Political Science and Public Policy at the University of Wisconsin–Madison. He has published numerous books and articles in IR in the fields of international political economy, international organizations, foreign policy analysis, and political methodology. He is a former editor of the leading IR field journal, International Organization.

Tanja A. Börzel
Freie Universität Berlin

Tanja A. Börzel is the Professor of political science and holds the Chair for European Integration at the Otto-Suhr-Institute for Political Science, Freie Universität Berlin. She holds a PhD from the European University Institute, Florence, Italy. She is coordinator of the Research College "The Transformative Power of Europe," as well as the FP7-Collaborative Project "Maximizing the Enlargement Capacity of the European Union" and the H2020 Collaborative Project "The EU and Eastern Partnership Countries: An Inside-Out Analysis and Strategic Assessment." She directs the Jean Monnet Center of Excellence "Europe and its Citizens."

Edward D. Mansfield
University of Pennsylvania

Edward D. Mansfield is the Hum Rosen Professor of Political Science, University of Pennsylvania. He has published well over 100 books and articles in the area of international political economy, international security, and international organizations. He is Director of the Christopher H. Browne Center for International Politics at the University of Pennsylvania and former program co-chair of the American Political Science Association.

Editorial Team

International Relations Theory
Jeffrey T. Checkel, European University Institute, Florence

International Security
Anna Leander, Graduate Institute Geneva

International Political Economy
Edward D. Mansfield, University of Pennsylvania
Stafanie Walter, University of Zurich

International Organisations
Tanja A. Börzel, Freie Universität Berlin
Jon C. W. Pevehouse, University of Wisconsin–Madison

About the Series

The Cambridge Elements Series in International Relations publishes original research on key topics in the field. The series includes manuscripts addressing international security, international political economy, international organizations, and international relations.

Cambridge Elements ᵀ

International Relations

Elements in the Series

Across Type, Time and Space: American Grand Strategy in Comparative Perspective
Peter Dombrowski and Simon Reich

Contestations of the Liberal International Order
Fredrik Söderbaum, Kilian Spandler, Agnese Pacciardi

Domestic Interests, Democracy, and Foreign Policy Change
Brett Ashley Leeds, Michaela Mattes

Token Forces: How Tiny Troop Deployments Became Ubiquitous in UN Peacekeeping
Katharina P. Coleman, Xiaojun Li

The Dual Nature of Multilateral Development Banks
Laura Francesca Peitz

Peace in Digital International Relations
Oliver P. Richmond, Gëzim Visoka, Ioannis Tellidis

Regionalized Governance in the Global South
Brooke Coe, Kathryn Nash

Digital Globalization
Stephen Weymouth

After Hedging: Hard Choices for the Indo-Pacific States Between the US and China
Kai He and Huiyun Feng

IMF Lending: Partisanship, Punishment, and Protest
Rodwan Abouharb, Bernhard Reinsberg

Building Pathways to Peace: State–Society Relations and Security Sector Reform
Nadine Ansorg and Sabine Kurtenbach

Drones, Force and Law: European Perspectives
David Hastings Dunn and Nicholas J. Wheeler with Jack Davies and Zeenat Sabur

A full series listing is available at www.cambridge.org/EIR

Printed in the United States
by Baker & Taylor Publisher Services